THE COMMONWEALTH AND INTERNATIONAL LIBRARY

Joint Chairmen of the Advisory Board

SIR ROBERT ROBINSON, O.M., F.R.S., LONDON
DEAN ATHELSTAN SPILHAUS, MINNESOTA

SCIENCE AND TECHNOLOGY

General Editor: G. E. BACON

AN INTRODUCTION TO
THE THEORY OF DIFFRACTION

Pergamon Materials Advisory Committee

AN INTRODUCTION TO THE THEORY OF DIFFRACTION

C. J. BALL

M.A., Ph.D., A.Inst.P.

Professor of Physics in the University of Zambia

PERGAMON PRESS

Oxford · New York · Toronto

Sydney · Braunschweig

Pergamon Press Ltd., Headington Hill Hall, Oxford
Pergamon Press Inc., Maxwell House, Fairview Park, Elmsford,
New York 10523
Pergamon of Canada Ltd., 207 Queen's Quay West, Toronto 1
Pergamon Press (Aust.) Pty. Ltd., 19a Boundary Street,
Rushcutters Bay, N.S.W. 2011, Australia
Vieweg & Sohn GmbH, Burgplatz 1, Braunschweig

First edition 1971

Library of Congress Catalog Card No. 75-151747

Printed in Great Britain by Bell & Bain Ltd., Glasgow

08 015786 6 (flexicover)
08 015787 4 (hard cover)

CONTENTS

PREFACE

DIFFRACTION techniques are increasingly being used in metallurgical studies, and there are already several very good books dealing in some detail with the theory and practice of X-ray, neutron and electron diffraction. Why, then, another book? It has been my experience in helping various groups of students, both undergraduate and postgraduate, that most difficulties arise from an unsure grasp of the fundamentals of diffraction theory, matters which are either too elementary to be dealt with in any detail in the books referred to above, or else thought to be too difficult to do more than state the results. I have, therefore, tried to fill these gaps in the literature. The emphasis throughout is on diffraction theory, rather than on applications, and so I have not attempted to discuss certain topics, such as absorption in cylindrical or slab-shaped specimens, which, though theoretical, do not closely involve the fundamentals of diffraction. Good accounts of these topics are to be found elsewhere. Also, I have sought after an understanding of diffraction and so have avoided, wherever possible, producing weird and wonderful formulae as it were out of a hat: memorising such formulae does nothing to help understanding.

The interest for metallurgists lies chiefly in three-dimensional diffraction (Chapter 4), but I make no apology for discussing also one- and two-dimensional diffraction (Chapters 2 and 3). Many of the fundamentals of diffraction theory can be introduced in this way, and more simply than in three dimensions. I have also included a chapter (Chapter 1) on the fundamentals of wave motion and a short account of the interaction of atoms with X-rays, neutrons and electrons (Chapter 5). This is designed merely to show the differences between atomic scattering factors for the different radiations: a true understanding of the interaction of matter with radiation is not to be had at this level. Chapter 6 is an introduction to Fourier methods, which I hope will make the student keen to learn more, and finally in Chapter 7 I have tried to allay some of the fears that are sometimes aroused by mention of the reciprocal lattice.

It would be surprising if there were much that is original in a book of this kind, and in this book there is not. I have drawn extensively on the books listed in the bibliography. Detailed references are not given in the text, with a few exceptions, but fuller accounts of most topics covered can be found in one or another of these books.

I am very grateful to Professor G. E. Bacon for reading the manuscript and making numerous helpful suggestions.

Lusaka, Zambia   C. J. BALL

CHAPTER 1

WAVE MOTION

1.1. Introduction

Most people are familiar with the appearance of waves on the sea or some other large body of water. Indeed, for the not too energetic, the fascination of watching a wave form at some distance from the shore and advance towards one never seems to fade. It is easy to concentrate attention on a single wave and think of it as an entity, in some way distinct from the sea, but if our attention should be distracted by a seagull bobbing up and down on the water we are reminded that the water in the wave is not advancing bodily towards the shore but instead is displaced only a relatively minor distance before returning to its original position as the wave passes. This is fundamental to the concept of a wave. The wave consists of a disturbance of the level of the water or, in more general terms, of the value of a physical quantity, the motion of the wave being quite distinct from the disturbance of the quantity. A scientist watching the waves will wonder about their origin and behaviour, advancing at just the rate they do and no other, eventually breaking on the shore. A first step in understanding the phenomenon is to describe it in mathematical terms. If we imagine a horizontal line drawn parallel to the direction of travel of the wave and represent distance along this line by the variable x, and let the vertical distance of the water surface from this line be represented by ϕ, then the equation of the line of intersection of the water surface with the vertical plane through the x-axis at some instant of time will be

$$\phi = f(x)$$

where $f(x)$ is some function that, in principle, we could determine if given sufficiently precise information (Fig. 1.1a). It will be convenient to choose our origin of ϕ such that $\phi = 0$ everywhere in the absence of the wave. The actual value of ϕ at any time and place during the passage of the wave is then known as the *displacement* at that time and place. At

1

some later instant of time the wave will have moved on and, very likely, have changed its shape slightly. We can, however, visualise an ideal wave in which the change of shape is so slight as to be negligible, in which case the same mathematical function as before will describe its shape provided that we choose a new origin of coordinates to allow for the distance that

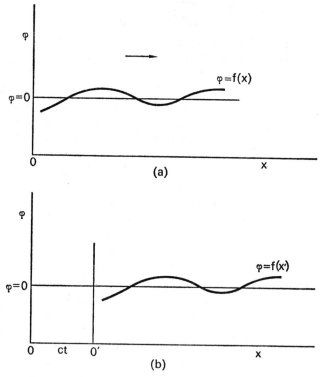

FIG. 1.1. Disturbance due to the passage of a wave: (a) at time $t = 0$; (b) at time $t = t$.

the wave has travelled (Fig. 1.1b). We will denote distance referred to the new origin by the variable x'. The line in Fig. 1.1b will be represented by the equation

$$\phi = f(x')$$

where $f(x')$ is the same function as before. If we choose our origin of time such that $t = 0$ in Fig. 1.1a and $t = t$ in Fig. 1.1b, then obviously $00' = ct$, where c is the velocity of the wave, and at any point on the x-axis

$$x' = x - ct.$$

The equation of the wave is therefore

$$\phi = f(x - ct) \tag{1.1}$$

when referred to a fixed system of coordinates. Had the wave been travelling in the opposite direction the form of the equation would have been

$$\phi = f(x + ct). \tag{1.2}$$

Differentiating either equation twice with respect to x and also with respect to t we find that

$$\frac{\partial^2 \phi}{\partial x^2} = \frac{1}{c^2} \cdot \frac{\partial^2 \phi}{\partial t^2}. \tag{1.3}$$

In other words, eqs. (1.1) and (1.2), which we know represent waves, are solutions of the differential eq. (1.3). The significance of this is that had we first concentrated our attention on the water, rather than on the wave, we would have arrived, with some approximations, at an equation similar in form to eq. (1.3). Indeed, in a very great variety of physical systems it can be shown that an equation identical in form with eq. (1.3) will be satisfied approximately by some physical quantity and in all such cases, therefore, we deduce that a disturbance in that quantity will propagate as a wave. Thus ϕ may represent the value of the electric field at a point in space (more exactly, one of its components), the displacement of a segment of a stretched string or the pressure in a gas. Equation (1.3) is known as the equation of wave motion in one dimension.

1.2. Parameters of a wave

An isolated wave can exist but it is far more usual to encounter a procession of waves, the wave profile repeating at regular intervals. Such a wave is known as a *periodic* wave. In this case we can define the *wavelength*, λ, as the distance between corresponding parts of neighbouring waves at any instant of time. We can also define the *frequency*, v, as the number of waves passing a fixed point in unit time. Obviously,

$$v\lambda = c.$$

We can also define the *wave number*, k, as the number of waves in unit length and the *period*, τ, as the time interval between the passage of corresponding parts of neighbouring waves. Then

$$k = 1/\lambda,$$

and

$$\tau = 1/v.$$

A particularly important class of waves is that in which the displacement can be represented by a simple sine or cosine function, e.g.

$$\phi = f(x-ct) = A\,\frac{\cos}{\sin}\left\{\frac{2\pi}{\lambda}(x-ct)-\delta\right\}. \tag{1.4}$$

Such a wave is known as an *harmonic* wave. For an harmonic wave we can further define the *amplitude*, A, as the maximum value attained by the displacement. The constant δ in eq. (1.4) allows for the fact that if the origins of x and t are chosen arbitrarily, then we will not in general have $\phi = A$ or 0 (for cos or sin waves) at $x = t = 0$. For a single wave we can always choose our origins such that this is the case, i.e. such that $\delta = 0$, but if more than one wave is involved we cannot ensure that this is the case for all the waves. δ is called the *phase* of the wave relative to a similar wave for which $\delta = 0$. Equation (1.4) can also be written in the form

$$\phi = A\,\frac{\cos}{\sin}\{2\pi(kx-vt)-\delta\},$$

which is often more convenient.

The *intensity* of a wave, I, is the energy density in the wave. For an harmonic wave the displacement at any point varies sinusoidally with time and the intensity is proportional to the square of the amplitude at that point, as for a simple harmonic oscillator. The intensity due to a source or sources of waves can and generally will vary from point to point in space and must not be confused with the rate of flow of energy in the waves.

1.3. Waves in three dimensions

The equation of wave motion in three dimensions is

$$\frac{\partial^2 \phi}{\partial x^2} + \frac{\partial^2 \phi}{\partial y^2} + \frac{\partial^2 \phi}{\partial z^2} = \frac{1}{c^2} \cdot \frac{\partial^2 \phi}{\partial t^2} \, . \tag{1.5}$$

Solutions can be found of the form

$$\phi = f(lx + my + nz - ct) \tag{1.6}$$

where $l^2 + m^2 + n^2 = 1$. The *wavefronts* are the surfaces on which ϕ has a constant value at any instance of time, i.e. the surfaces for which $lx + my + nz = $ const. These surfaces are parallel planes whose common normal has the direction cosines l, m, n, and so eq. (1.6) represents a plane wave in three dimensions. An harmonic plane wave can be represented by

$$\phi = A \cos\left\{\frac{2\pi}{\lambda}(lx + my + nz - ct) - \delta\right\}$$

$$= A \cos\{2\pi(k_1 x + k_2 y + k_3 z - vt) - \delta\}$$

$$= A \cos\{2\pi(\mathbf{k} \cdot \mathbf{r} - vt) - \delta\}$$

where $\mathbf{r}$ is the vector from the origin to the point (x, y, z) and $\mathbf{k}$ is the vector with components $(k_1, k_2, k_3) = (l/\lambda, m/\lambda, n/\lambda)$. Obviously $\mathbf{k}^2 = 1/\lambda^2$, i.e. $|\mathbf{k}| = 1/\lambda$. $\mathbf{k}$ is known as the *wave vector* of the wave.

There are other solutions not of this form, the most important being those representing spherical waves. To simplify discussion of these, eq. (1.5) can be written in spherical polar coordinates (Fig. 1.2)

$$\frac{\partial^2 \phi}{\partial r^2} + \frac{2}{r} \cdot \frac{\partial \phi}{\partial r} + \frac{1}{r^2 \sin\theta} \cdot \frac{\partial}{\partial \theta}\left\{\sin\theta \frac{\partial \phi}{\partial \theta}\right\} + \frac{1}{r^2 \sin^2\theta} \cdot \frac{\partial^2 \phi}{\partial \psi^2} = \frac{1}{c^2} \cdot \frac{\partial^2 \phi}{\partial t^2} \, . \tag{1.7}$$

If ϕ is independent of θ and ψ this reduces to the simpler equation

$$\frac{\partial^2 \phi}{\partial r^2} + \frac{2}{r} \cdot \frac{\partial \phi}{\partial r} = \frac{1}{c^2} \cdot \frac{\partial^2 \phi}{\partial t^2} \, ,$$

which can be written

$$\frac{\partial^2}{\partial r^2}(r\phi) = \frac{1}{c^2} \cdot \frac{\partial^2}{\partial t^2}(r\phi),$$

from which it is apparent that it has solutions of the form

$$r\phi = f(r-ct),$$

i.e.

$$\phi = \frac{1}{r}f(r-ct). \qquad (1.8)$$

The wavefronts are the surfaces $r = $ const, i.e. spheres, so eq. (1.8)

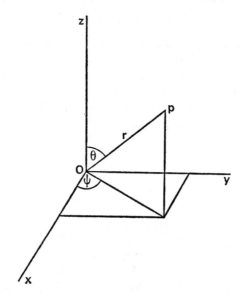

FIG. 1.2. Relation of spherical polar coordinates r, θ and ψ to Cartesian coordinates x, y and z.

represents a spherical wave. An harmonic spherical wave can be represented by

$$\phi = \frac{a}{r}\cos\{2\pi(kr-vt)-\delta\}. \qquad (1.9)$$

For sufficiently large values of r the variation of a/r will be much less rapid than that of the cosine term and so, in a limited region of space, the spherical waves will approximate to plane waves of constant amplitude with a direction of propagation parallel to OP, as shown in Fig. 1.3. This approximation is equivalent to neglecting the second term in eq. (1.7). Other solutions exist which approximate to eq. (1.9) at large values of r but in which a is not a constant independent of direction but is a function of θ and ψ. The best known of these is that describing the radiation from an oscillating dipole of very small size, for which $a = a_0 \sin \theta$ if the direction of the dipole is taken as the direction $\theta = 0$.

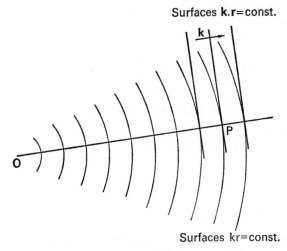

Surfaces **k.r**=const.

Surfaces kr=const.

FIG. 1.3. Approximation of a spherical wave to a plane wave.

1.4. The principle of superposition

The equation of wave motion, eq. (1.5), is a linear differential equation, i.e. it contains terms that are constant multiples of a function (ϕ) and the functions obtained from that function by differentiation. It does not contain terms that are the product of two or more of these functions, nor does it contain any terms involving these functions to other than the first power. An important mathematical consequence of this is that solutions of the equation of wave motion are additive—that is, if

$$\phi_1 = f_1(x-ct)$$

and

$$\phi_2 = f_2(x-ct)$$

are two solutions, then

$$\phi = \phi_1 + \phi_2$$

is also a solution, as can easily be seen by substitution. Obviously the sum of any number of solutions will itself be a solution. This is one statement of the principle of superposition. It can be restated in the form "the resultant displacement at any point in space due to the passage of any number of waves is the sum of the displacements that would be produced by the individual waves acting alone". In essence, all problems in diffraction consist of first determining what waves will pass through a given point and then performing the necessary summation.

1.5. Amplitude–phase diagrams

Consider an harmonic spherical wave originating at a source O (Fig. 1.3). The equation of this wave will be

$$\phi_1 = \frac{a_1}{r} \cos \{2\pi(kr - vt) - \delta_1\}.$$

This can also be written as

$$\phi_1 = \mathcal{R}[A_1 \exp(\pm i\{2\pi(kr - vt) - \delta_1\})]$$

where $\mathcal{R}$ signifies that ϕ is the real part of the complex quantity in square brackets and $A_1 = a_1/r$. This will henceforth be written simply as

$$\phi_1 = A_1 \exp[+i\{2\pi(kr - vt) - \delta_1\}] \tag{1.10}$$

where it is understood that ϕ is equal only to the real part of the complex quantity and we have arbitrarily chosen the plus sign in the exponential.

The convenience of the exponential representation of ϕ lies in the fact that the space and time coordinates can easily be separated, thus eq. (1.10) can be written

$$\phi_1 = A_1 e^{i\alpha_1} \cdot e^{-2\pi ivt} \tag{1.11}$$

where $\alpha_1 = 2\pi kr - \delta_1$. The quantity $A_1 e^{i\alpha_1}$ involving both the amplitude and the phase angle, α_1, is known as the *complex amplitude*.

The complex number on the right of eq. (1.11) can be represented by a point z_1 in the complex plane, whose modulus is A_1 and whose argument is

$(\alpha_1 - 2\pi v t)$ (Fig. 1.4). As t increases, z_1 describes a circle about the origin, completing one revolution for an increase in t of v^{-1}. If a second wave is producing a displacement given by

$$\phi_2 = A_2 e^{i\alpha_2} \cdot e^{-2\pi i v t}$$

the resultant displacement at any point will be given by

$$\phi = \phi_1 + \phi_2 = (A_1 e^{i\alpha_1} + A_2 e^{i\alpha_2}) \cdot e^{-2\pi i v t}$$

provided that the physical quantity represented by ϕ is a scalar or, if a vector, that ϕ_1 and ϕ_2 are parallel.

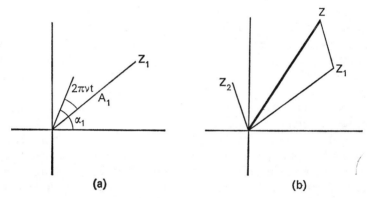

(a)   (b)

FIG. 1.4. Amplitude–phase diagrams.

As t increases the point $z = z_1 + z_2$ (Fig. 1.4b) will describe a circle in the complex plane. The displacement at any time will be the real part of z, but the amplitude will be its modulus, and this will be the modulus of the complex amplitude $(A_1 e^{i\alpha_1} + A_2 e^{i\alpha_2})$. Since the intensity depends only on the amplitude and not on the value of the displacement at any specified time, it is not necessary to consider the time factor $e^{-2\pi i v t}$ explicitly. The amplitude of the disturbance at a point is simply the modulus of the complex number obtained by adding the complex amplitudes $A_j e^{i\alpha_j}$ for each of the waves $(j = 1, 2, 3, ...)$ incident on that point.

A diagram such as Fig. 1.4b, showing addition of complex amplitudes for a number of waves, is known as an amplitude–phase diagram. Since rotation of the diagram about the origin is of no significance we may choose one of the phase angles arbitrarily. This is equivalent to choosing an origin of time. It is usual, but by no means necessary, to choose the phase angle of the first wave considered to be zero.

1.6. Secondary wavelets

When considering the origin of the several waves that combine to produce a diffraction pattern it is helpful to distinguish between fictitious and physical scattering processes. As an example of the former, Fresnel, extending a theory due to Huygens, postulated that each element of surface of a wavefront in a beam of light acts as a source of spherical waves, the amplitude of the wave at unit distance from the element being proportional to its area. This is known as Huygens' principle. It was also necessary to postulate, as we shall see later, that the phase of the secondary wavelets is $\pi/2$ ahead of that of the incident wave and that their amplitude is inversely proportional to the wavelength of the light and decreases with increasing angle from the original direction of propagation of the wave. Kirchhoff subsequently put the theory on a sound basis by showing, from a consideration of the equation of wave motion, that the displacement at any point could be calculated from a knowledge of ϕ, $\partial\phi/\partial t$ and $\partial\phi/\partial n$ (where $\partial\phi/\partial n$ denotes differentiation along the normal to the surface) at all points on a closed surface inside which ϕ and its derivatives are finite and continuous. Furthermore, the integral from which ϕ is calculated is the same as would be obtained by applying Fresnel's theory. Kirchhoff's integral includes the $\pi/2$ phase factor and the amplitude factors: the decrease in amplitude with angle θ from the original direction of propagation, the obliquity factor, is shown to be proportional to $(1+\cos\theta)$. For a derivation of the Kirchhoff diffraction integral see Longhurst (1957).

It needs to be emphasised that the Kirchhoff diffraction integral is a mathematical consequence of the equation of wave motion. It does not imply that there is any physical process by which the original wave is scattered and it is equally valid for any physical quantity obeying the equation of wave motion. One important difference between this case and cases in which physical scattering does take place is that, when applying Huygens' principle, the original wave is totally replaced by the secondary wavelets, whereas when considering physical scattering it is not and both the original wave and any scattered waves must be considered. A second important difference is that, very often, physical scattering processes are incoherent—that is, the scattered waves are not of the same wavelength as the incident wave and scattered waves from different parts of a specimen, even when of the same wavelength, do not bear a constant phase relation to one another. Some examples of physical scattering processes are given in Chapter 5.

CHAPTER 2

ONE-DIMENSIONAL DIFFRACTION

2.1. Introduction

In this chapter we will consider a number of problems in which the distribution of scattering sources is either one-dimensional in nature or can be reduced to one dimension.

Diffraction phenomena are customarily classified as being either "Fresnel" or "Fraunhofer" in character. The distinction is useful, but this is not to say that there is any substantial difference between the two categories of phenomena: the principles and methods of treatment are the same in both cases. In the most usual arrangement for observing Fraunhofer diffraction in optics (see Fig. 2.1), the wave incident on the diffracting screen is plane and the waves reaching any point in the plane of observation are parallel (or nearly so), after being scattered. Fresnel diffraction, on the other hand, is usually observed in planes close to a diffracting aperture (or obstacle), i.e. at distances that are not very much larger than the dimensions of the aperture. Mathematically, the significant difference between the two cases is that in the former the phase difference between waves originating from two scattering centres is simply proportional to their separation whereas in the latter it is not, depending on the positions of the scattering centres within the aperture as well. Alternatively, if the phase differences between waves from different scattering centres and a fixed origin are expressed in terms of the distances of the centres from the origin, then for Fraunhofer diffraction the phase difference will depend on the first power of the distance only, whereas in Fresnel diffraction the expression for the phase difference will include terms in higher powers of the distance as well. As might be expected, in practice classification is not always clear-cut. Often, in comparison with the first-order term the higher-order terms are small, but not zero: whether they are negligible or not depends on the desired accuracy of calculation. This makes it clear that Fraunhofer diffraction should be thought of as a limiting case of the more general Fresnel phenomenon.

The arrangement of Fig. 2.1 is not the only arrangement for observing Fraunhofer diffraction. It can be shown that the condition that the phase

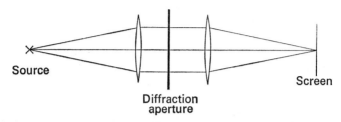

FIG. 2.1. Usual arrangement for observing Fraunhofer diffraction (schematic).

difference between waves from two centres be proportional to their separation will be satisfied if the diffraction pattern is viewed in a plane conjugate to the light source, i.e. the pattern is an image of the light source, formed by an arrangement of lenses or mirrors and modified by an intermediate aperture. It will also be satisfied approximately if the source and plane of observation are at a great distance from the diffraction aperture or scattering centres, which is almost invariably the case when dealing with X-ray or neutron diffraction.

2.2. The double source

Consider two scattering centres S_1 and S_2, a distance d apart, giving rise to scattered waves

$$\phi_1 = \frac{a_1}{r_1} \exp \left[i\{2\pi k r_1 - \delta_1\} \right] . \exp \left(-2\pi i v t \right), \quad \text{etc.,}$$

and suppose that the diffraction pattern is viewed along a line AA' parallel to S_1S_2 and at a distance D from S_1S_2 where $D \gg d$, as in Fig. 2.2.

At any point such as P waves from S_1 and S_2 will have travelled distances S_1P and S_2P which differ by

$$S_2P - S_1P = S_2T_2 + S_1T_1$$

$$\simeq d . \sin \theta$$

$$\simeq d . x/D.$$

The approximations are valid provided that $d \ll D$ and θ is small.

If S_1 and S_2 are scattering in phase with one another ($\delta_1 = \delta_2$), the phase difference between the waves arriving at P will simply be due to the unequal paths travelled and will be

$$\varepsilon = \alpha_2 - \alpha_1 = \frac{2\pi}{\lambda} \frac{d}{D} x.$$

The resultant amplitude at P can easily be found from an amplitude–phase diagram, Fig. 2.3, in which α_1 has been taken as zero.

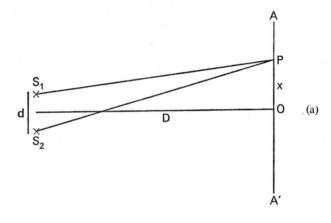

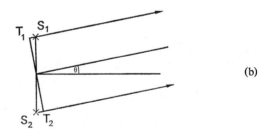

Fig. 2.2. Geometry of diffraction pattern produced by two sources radiating in phase. (a) d greatly exaggerated with respect to D; (b) detail of waves near sources.

Since $r_2 = r_1$ and the obliquity factors for the two waves are also very closely the same, $A_2 = A_1$ and so

$$A(x) = |\phi| = 2A_1 \cos\left(\frac{\pi \, dx}{\lambda D}\right).$$

The intensity at P will be given by

$$I(x) = A(x)^2 = 4A_1^2 \cos^2\left(\frac{\pi \, dx}{\lambda D}\right).$$

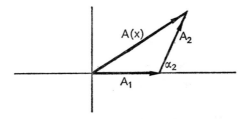

FIG. 2.3. Amplitude–phase diagram for double source.

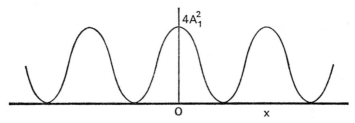

FIG. 2.4. Intensity produced on screen by arrangement of Fig. 2.2.

The intensity distribution along the line AA' on the screen will therefore be as shown in Fig. 2.4. The maximum intensity is $4A_1^2$, the minimum zero, and the average intensity is equal to the sum of the intensities that would be produced by S_1 and S_2 separately.

If the sources S_1 and S_2 are not radiating in phase, but the phase difference between them is constant, the maxima of Fig. 2.4 will be displaced to right or left by an amount depending on the phase difference. The intensity distribution between maxima will, of course, be the same as before.

If the phase difference between S_1 and S_2 is not constant but is varying randomly in the range $0 \leqslant$ phase difference $\leqslant 2\pi$, then during any short interval of time, during which the phase difference does not change by much, the intensity distribution will be as shown in Fig. 2.4, suitably shifted. During a later interval of time the intensity distribution will again be as in Fig. 2.4, but with the maxima occurring at different places. During a long time interval, then, the average intensity at any place will be equal to the average value of the intensity distribution of Fig. 2.4, and will be the same at all places. The average intensity observed at any point on the screen will therefore be equal to the sum of the intensities that would have been produced by the two sources acting separately.

This result can be generalised to any number of sources: if the relative phases of a number of sources are completely random, the time average of the intensity produced at any point will be equal to the sum of the intensities that would have been produced by the individual sources acting separately.

If the relative phases are not completely random, then the intensity distribution is not so easily calculated; for two sources it will be the weighted average of intensity curves such as that of Fig. 2.4, displaced by amounts depending on the phase differences and weighted according to the probabilities of those phase differences occurring.

One special case is of some interest, where the phase difference is not completely random but can take only the values 0 or π, with equal probability. The weighted average of the intensity curves will again be the same at all points, so the time average of the intensity at any point is the same as it would have been had the phases been completely random, i.e. it is obtained by summing intensities.

2.3. The multiple source

Consider N scattering centres that are uniformly spaced on a straight line with a distance d between each pair. If the diffraction pattern is viewed at a great distance or in the focal plane of a lens system, i.e. under Fraunhofer conditions, the phase difference between each pair of adjacent scattered waves will be the same, i.e.

$$\varepsilon = \frac{2\pi}{\lambda} d \sin \theta.$$

The amplitude–phase diagram for this case is shown in Fig. 2.5, where the phase of the first wave has again been taken as zero. The resultant amplitude is obtained most easily by addition of the complex amplitudes, thus

$$A = A_1 e^{i\alpha_1} + A_2 e^{i\alpha_2} + \ldots + A_N e^{i\alpha_N}$$
$$= A_1 + A_2 e^{i\varepsilon} + A_3 e^{2i\varepsilon} + \ldots + A_N e^{(N-1)i\varepsilon}.$$

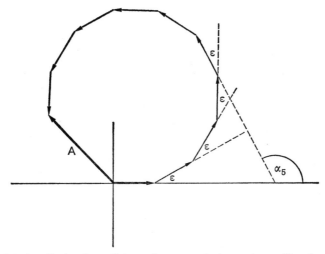

FIG. 2.5. Amplitude–phase diagram for ten scattering centres uniformly spaced on a straight line.

Now $A_1 = A_2 = \ldots = A_N$,

$$\therefore \qquad A = A_1 \cdot \frac{e^{Ni\varepsilon} - 1}{e^{i\varepsilon} - 1}$$

$$= A_1 \cdot \frac{e^{\frac{1}{2}Ni\varepsilon}}{e^{\frac{1}{2}i\varepsilon}} \cdot \frac{e^{\frac{1}{2}Ni\varepsilon} - e^{-\frac{1}{2}Ni\varepsilon}}{e^{\frac{1}{2}i\varepsilon} - e^{-\frac{1}{2}i\varepsilon}}$$

$$= A_1 \cdot e^{\frac{1}{2}(N-1)i\varepsilon} \cdot \frac{\sin\left(\frac{1}{2}N\varepsilon\right)}{\sin\left(\frac{1}{2}\varepsilon\right)}.$$

The factor $e^{\frac{1}{2}(N-1)i\varepsilon}$ simply expresses the fact that the phase of the resultant relative to that of the first wave is $\frac{1}{2}(N-1)\varepsilon$. The amplitude of the resultant is the modulus of the complex amplitude, i.e.

$$|A| = A_1 \frac{\sin (N\pi d \sin \theta/\lambda)}{\sin (\pi d \sin \theta/\lambda)} . \tag{2.1}$$

The resulting intensity is the square of this expression. It is plotted as a function of $(\pi d \sin \theta/\lambda)$ in Fig. 2.6 for the case $N = 10$.

The principal maxima occur at angles such that $\gamma \equiv (\pi d \sin \theta/\lambda) = m\pi$, where m is any integer. The value of I at the principal maxima is just $I_0 = N^2 A_1^2$, since all the scattered waves are in phase and the resultant amplitude is NA_1. Secondary maxima occur close to the angles for which $(N\pi d \sin \theta/\lambda) = (n+\frac{1}{2})\pi$, with $n = \pm 1, \pm 2, \ldots$, but excluding $n = Nm$

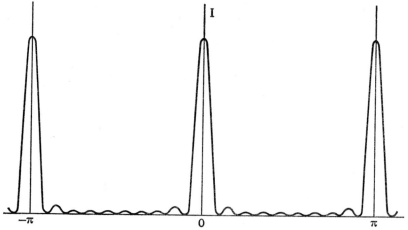

Fig. 2.6. Fraunhofer intensity distribution produced by ten scattering centres, as a function of $\pi d \sin \theta/\lambda$.

(which angles fall inside a principal maximum) provided that N is so large that the variation of the denominator of eq. (2.1) is much less rapid than that of the numerator. The first secondary maximum on either side of a principal maximum will have an intensity of

$$I_1 = \frac{A^2}{\sin^2 (3\pi/2N)}$$

$$\simeq N^2 A^2 \cdot \frac{4}{9\pi^2}$$

$$= 0.045 \, I_0 .$$

Note that this is independent of N.

Minima occur at angles such that $(N\pi d \sin \theta/\lambda) = n\pi$, with $n = \pm 1$, ± 2, ..., but excluding $n = Nm$, which angles correspond to principal maxima. The minima adjacent to the principal maximum of order m are at angles θ such that

$$\sin \theta = \frac{\lambda}{d}\left(m \pm \frac{1}{N}\right).$$

2.4. Diffraction by a slit aperture

So far we have considered point scattering centres but we have not made any assumptions concerning the angular dependence of the amplitude of the scattered waves. In Chapter 5 we will consider this problem for the scattering of X-rays, electrons and neutrons by atoms: we will now consider the diffraction of a plane, semi-parallel beam of light by a slit aperture, under Fraunhofer conditions. By semi-parallel is meant a beam in which the divergence is very small in one plane but large in a plane at right angles to this plane. Such a beam would be produced by illuminating a slit in the focal plane of a lens: the divergence is large in the plane containing the length of the slit but small in a plane normal to the slit. We suppose that the source slit is parallel to the diffracting slit.

In order to apply Huygens' principle to this problem we need to know the amplitude and phase of the incident wave at all points on a surface bridging the aperture, which we may conveniently take in the plane of the aperture. In general these are not known but as a very good first approximation in cases where the aperture is large in comparison with the wavelength we may suppose that they are the same as they would be in the absence of the blocking screen. (This assumption is known as St. Venant's principle.) That this cannot be the case exactly is apparent from the fact that it implies that ϕ is discontinuous at the boundary of the aperture.

The problem, strictly, is one in two dimensions. We have to sum the amplitudes of the secondary wavelets from all elements of area of the aperture. It can be reduced to a problem in one dimension by dividing the aperture into identical narrow strips parallel to the length of the aperture. The phase of the incident wave will vary continuously in a random manner along the length of a strip but the variation will be the same in all strips. Because of the lack of phase correlation along the strip, radiation from it will not be confined to directions in the plane normal to it, and so the strip may be replaced by a single scattering centre with scattering power proportional to its width. The resultant amplitude diffracted by the aperture in a direction in the plane normal to the length of the aperture and making an angle θ with the incident beam direction in this plane can therefore be obtained from eq. (2.1) by letting the number of strips go to infinity, at the same time making allowance for the fact that the amplitude of the wave scattered by each strip and also the distance between neighbouring strips will be proportional to N^{-1}.

If we put $A_1 = A_0/N$, where A_0 is the resultant amplitude when all the waves are in phase, and $d = a/N$, where a is the width of the aperture, then eq. (2.1) becomes

$$|A| = \frac{A_0}{N} \cdot \frac{\sin (\pi a \sin \theta/\lambda)}{\sin \left(\pi \dfrac{a}{N} \dfrac{\sin \theta}{\lambda} \right)},$$

$$\therefore \lim_{N \to \infty} |A| = \frac{A_0 \sin \beta}{\beta} \tag{2.2}$$

where

$$\beta = \frac{\pi a \sin \theta}{\lambda}.$$

Note that β is half the phase difference between waves from opposite sides of the aperture.

A_0 will also vary with θ, on account of the obliquity factor, but for most apertures A_0 may be taken as constant since $\beta^{-1} \sin \beta$ will have fallen to negligible values, even at the maxima, before there is any appreciable decrease in A_0. Thus, for a slit of width 0·1 mm and for light of wavelength 5×10^{-5} cm (green) the intensity of the secondary maxima will have fallen to 0·1 % of that of the principal maximum for a value of $\beta \sim 32$, i.e. at an angle θ such that $\sin \theta = \lambda \beta/\pi a \simeq 5 \times 10^{-2}$. At this angle A_0 will have decreased by less than 0·1 %. However, for slits of width comparable to the wavelength of the radiation used, as can easily happen with microwaves, the obliquity factor would be important, but under these conditions the error involved in applying St. Venant's principle would also not be negligible.

The intensity of the diffracted wave in the plane normal to the length of the aperture is shown in Fig. 2.7a as a function of β. The amplitude–phase diagram is shown in Fig. 2.7b, where the phase of the wavelet from the edge of the strip has been taken as zero. Equation (2.2) can readily be derived from this diagram if we remember that A_0 is the length of the arc, and is constant.

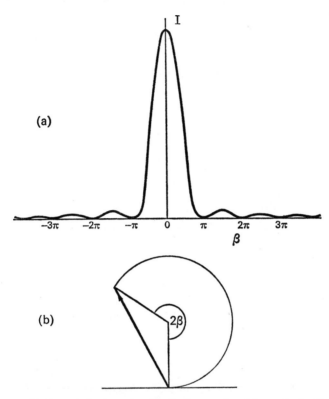

Fig. 2.7. (a) Fraunhofer intensity distribution produced by a single slit, as a function of β; (b) the corresponding amplitude–phase diagram.

The full diffraction pattern, which is two-dimensional, will consist of a series of lines parallel to the source slit, the intensity profile along a line normal to the lines at any height being that of Fig. 2.7a.

Diffraction by a slit aperture under Fresnel conditions can be treated by considering the aperture to be two parallel straight edges in close proximity. We will now consider diffraction by a single straight edge under Fresnel conditions.

2.5. Diffraction at a straight edge

2.5a. *Theory*

In Fig. 2.8 an opaque screen lies in the half-plane $x = 0$, $y < 0$, and has one edge lying along the z-axis. A plane wave is travelling in the x direction and is incident on the screen from the side $x < 0$. We wish to calculate the intensity at some point $P\ (x, y, z)$, which we may without loss of generality take in the plane $z = 0$, since the screen extends from $-\infty < z < \infty$.

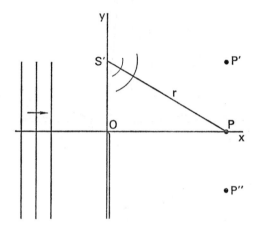

Fig. 2.8. Geometry of diffraction by a straight edge.

In order to do this we must sum the amplitudes of the secondary waves from all elements of area in the half-plane $x = 0$, $y > 0$. As a first step we divide the half-plane into identical narrow strips parallel to the z-axis. Each strip will act as the centre of a cylindrical wave and so, for a consideration of the diffraction pattern in the plane $z = 0$, may be replaced by a point source on the y-axis of scattering power proportional to its width, dy. This reduces the problem to a one-dimensional summation. We do not assume that the phase of the cylindrical wave is the same as the phase of the secondary wavelet from the element of area at the centre of the strip (in fact it is not), but merely that the difference of phase is the same for all strips.

Consider first the intensity at a point on the x-axis. The wave originating at S' (Fig. 2.8) will have travelled a distance $r = (x^2+y^2)^{\frac{1}{2}}$. Its phase relative to that of a wave from O will be

$$2\pi k(r-x)$$

$$= 2\pi k\left\{x\left(1+\frac{y^2}{x^2}\right)^{\frac{1}{2}}-x\right\}$$

$$\simeq \frac{\pi k}{x}y^2$$

where terms in $(y^2/x^2)^2$ and higher powers have been neglected. To this approximation the obliquity factor and the difference between $1/r$ and

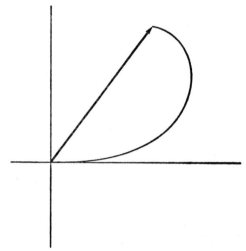

FIG. 2.9. Partial amplitude–phase diagram for the disturbance at P in Fig. 2.8.

$1/x$ can also be neglected, so the amplitude of the wave is simply proportional to dy, the width of the strip.

It is convenient to transform to the dimensionless variable $s = (2k/x)^{\frac{1}{2}}y$. When this is done the phase of the wave from S' becomes

$$\alpha = \tfrac{1}{2}\pi s^2$$

and its amplitude becomes

$$dA = A'ds$$

where A' remains to be determined. Part of the amplitude–phase diagram for the half-plane is shown in Fig. 2.9.

If A' is independent of s (i.e. the distance and obliquity factors are negligible), the equation of the amplitude–phase diagram is

$$X = A' \int_0^s \cos \left(\tfrac{1}{2}\pi s^2\right) ds,$$

$$Y = A' \int_0^s \sin \left(\tfrac{1}{2}\pi s^2\right) ds$$

since $dX = dA \cos \alpha$ and $dY = dA \sin \alpha$.

These integrals are known as Fresnel's integrals. Their evaluation is a simple matter with the aid of a digital computer. A table of the Fresnel integrals can be found in Jenkins and White (1950).

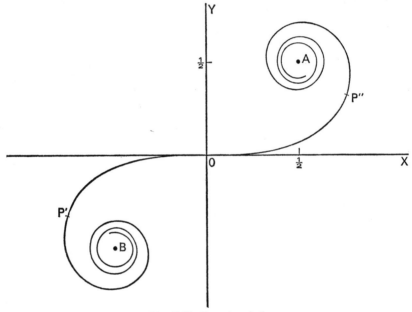

FIG. 2.10. Cornu's spiral.

The graphical representation of the Fresnel integrals is known as Cornu's spiral. It should be noted that Cornu's spiral is only an approximation to the true amplitude–phase diagram on account of the approximations made in its derivation. These, however, introduce negligible error if $\lambda \ll x$.

The part of the wave front that was obstructed by the screen would have contributed an identical amplitude–phase diagram, only inverted through the origin, so that the full amplitude–phase diagram is as shown in Fig. 2.10.

The limits of the Fresnel integrals as $s \to \infty$ are $X = Y = \frac{1}{2}$. The resultant amplitude from an unobstructed wavefront is therefore $A = \sqrt{2} \cdot A' = A_0$, and so $A' = A_0/\sqrt{2}$, where A_0 is the amplitude of the incident wave.

Cornu's spiral may be used to calculate the intensity at any point, on or off the axis, as follows.

For a point such as P' (Fig. 2.8), outside the geometric shadow, we choose the origin of phase as the phase of the element at S', nearest to P'. The elements above S' in Fig. 2.8 will then contribute a resultant amplitude OA. The amplitude phase diagram for those elements below S' will be a segment OP' of the Cornu spiral. The resultant amplitude at P' is therefore $P'A$. P' in Fig. 2.10 can be located by evaluating the corresponding value of s from the geometry of Fig. 2.8 and the wavelength of the light used.

For a point P'' inside the geometrical shadow the procedure is the same, but now the only part of the wavefront to contribute is that corresponding to the segment $P''A$ of the spiral. $P''A$ is therefore the resultant amplitude.

To deduce the Fresnel diffraction patterns of slit apertures or obstacles, the relevant section of the spiral can be obtained by evaluating s_1 and s_2 for the elements at each end of the aperture or obstacle. For an aperture the resultant amplitude is then the vector joining the two points s_1 and s_2 on the spiral, whilst for an obstacle the resultant amplitude is the vector $Bs_1 + s_2A$. In practice this would be done not graphically but from a table of the Fresnel integrals, as in the following example.

2.5b. *Example*

We will calculate the intensity of the light in a plane at a distance $x = 0 \cdot 1$ mm behind the edge of an opaque screen when it is illuminated with a beam of light of wavelength 5×10^{-5} cm, as in Fig. 2.8. For this case

$$
\begin{aligned}
s &= (2k/x)^{\frac{1}{2}}y \\
&= (2 \times 2 \times 10^4/10^{-2})^{\frac{1}{2}}y \\
&= 2 \times 10^3 y \qquad (y \text{ in cm}) \\
&= 0 \cdot 2y \qquad (y \text{ in } \mu\text{m}).
\end{aligned}
$$

The calculation of the intensity at various values of y in the plane $x = 0 \cdot 1$ mm, for an incident intensity $I_0 = 2 \, (A' = 1)$, could be set out as

B

Theory of Diffraction

in Table 2.1. Each line of the table show the stages in the calculation of the intensity at one value of y. The points in the table are plotted in Fig. 2.11.

Straight edge Fresnel diffraction patterns are frequently observed in microscopy (optical or electron) when the microscope is incorrectly focused.

TABLE 2.1

y (μm)	s_P	X_P	Y_P	$(PA)_X$	$(PA)_Y$	I
−10	2·0	0·49	0·34	0·01	0·16	0·03
−6	1·2	0·72	0·62	−0·22	−0·12	0·06
−2	0·4	0·40	0·03	0·10	0·47	0·23
0	0·0	0·00	0·00	0·50	0·50	0·50
2	0·4	−0·40	−0·03	0·90	0·53	1·09
4	0·8	−0·72	−0·25	1·22	0·75	2·04
6	1·2	−0·72	−0·62	1·22	1·12	2·73
8	1·6	−0·37	−0·64	0·87	1·14	2·06
10	2·0	−0·49	−0·34	0·99	0·84	1·69
12	2·4	−0·56	−0·62	1·06	1·12	2·37
14	2·8	−0·47	−0·40	0·97	0·90	1·75

Notes

1. P is in the positive quadrant of Fig. 2.10 if in the geometrical shadow of Fig. 2.8, hence the difference in sign between y and (X_P, Y_P).

2. X_P and Y_P are taken from tables of the Fresnel integrals (cf. Jenkins and White, 1950).

3. $(PX)_X$, the X component of PA, $= 0.5 - X_P$. Similarly, $(PA)_Y = 0.5 - Y_P$.

4. $I = (PA)_X{}^2 + (PA)_Y{}^2$.

2.6. The diffraction grating

2.6a. *Angular distribution of intensity*

A diffraction grating generally consists of a plastic replica of a glass or metal surface on which a large number of fine, parallel, evenly spaced lines have been ruled. It approximates to a series of narrow slit apertures, the irregular thickness of the plastic that filled the scratches serving to scatter the light incident on these regions and the uniform thickness of the plastic covering the untouched regions acting as transparent slits.

The diffraction grating can be treated as an array of N identical scattering centres, the angular dependence of the amplitude of the wave scattered from each centre being that of a finite slit. The angular distribution of intensity when the grating is illuminated at normal incidence with a plane parallel beam of light can therefore be obtained by combining the results of Sections 2.3 and 2.4, and is

$$I = I_0 \frac{\sin^2 \beta}{\beta^2} \frac{\sin^2 (N\gamma)}{\sin^2 \gamma} \tag{2.3}$$

in the same notation as before.

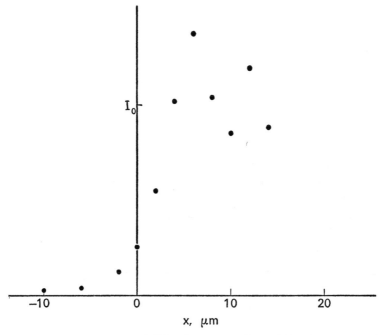

Fig. 2.11. Fresnel diffraction pattern of a straight edge.

The variation with θ of $\sin^2 (N\gamma)/\sin^2 \gamma$ is very much more rapid than that of $\sin^2 \beta/\beta^2$ and so the maxima of I are almost exactly at the angles of the maxima of $\sin^2(N\gamma)/\sin^2 \gamma$, i.e. principal maxima at angles such that $\pi d \sin \theta/\lambda = m\pi$, or

$$\sin \theta = m\lambda/d.$$

The intensities of the principal maxima are determined by the single-slit diffraction pattern; in effect we are able to determine the intensity in the single slit diffraction pattern only at those angles that satisfy eq. (2.4). For a grating with 15,000 lines per inch and for a wavelength of 500 nm, $d = 3\cdot3\lambda$, so the maximum order of diffraction that can be observed is $m = 3$. It is not surprising, therefore, that we are unable to infer in great detail the characteristics of a single slit from observations of the diffraction pattern. We will return to this subject in Section 2.8.

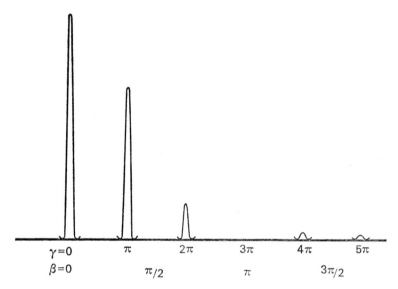

FIG. 2.12. Principal maxima in Fraunhofer intensity distribution for a diffraction grating of ten lines with $d = 3a$.

The angular distribution of intensity for a grating with ten lines for which $d = 3a$ is shown in Fig. 2.12. Note that the width of each principal maximum is the same. For a grating with a large number of lines, for which the intensity of the subsidiary maxima falls to a small fraction of the intensity of the principal maxima quite close to the latter, the intensity distribution around each of the principal maxima is the same as in the diffraction pattern of a single slit of width equal to the total width of the grating. This result can be obtained from eq. (2.3) by putting $I_0 = 1/N^2$, to give a zero order maximum of unit intensity, and approximating $\sin^2 \gamma$ by γ^2. Equation (2.3) then becomes

$$I = \frac{\sin^2 \beta}{\beta^2} \frac{\sin^2 (N\gamma)}{(N\gamma)^2}.$$

The second term gives the angular distribution of intensity from a single slit of width (Nd), i.e. the width of the grating.

2.6b. *Oblique illumination*

If a diffraction grating is illuminated with a plane parallel beam of light at other than normal incidence, as in Fig. 2.13, an additional phase

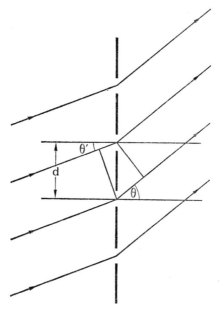

FIG. 2.13. Geometry of path difference when a grating is illuminated at oblique incidence.

difference between the wavelets from different slits is introduced since these wavelets no longer start in phase with each other. The phase difference between wavelets from adjacent slits becomes

$$\varepsilon = \frac{2\pi d}{\lambda} (\sin \theta - \sin \theta').$$

With this modification the analysis is the same as before, so eq. (2.3) still applies but with sin θ replaced by (sin θ – sin θ') in the expressions for β and γ. If θ' is small the effect is simply to displace the diffraction pattern by an angle $\theta'/\cos\theta$ to higher or lower values of θ, depending on the sign of θ'.

2.6c. *Resolving power*

The resolving power of a grating is a measure of its ability to show as separate two spectral lines in the same source whose wavelengths are nearly the same. It is defined as the ratio $\lambda/\Delta\lambda$, where $\Delta\lambda$ is the smallest wavelength difference that can be distinguished at a wavelength λ. In practice this will depend on the conditions of illumination and the visual acuity of the observer as well as on the properties of the grating. It is therefore the practice to adopt an arbitrary criterion of resolution (Rayleigh's criterion). According to this, two spectral lines will appear to be just resolved if a principal maximum of one falls on the first minimum of the other when the grating is illuminated with a strictly parallel beam of light. This implies that

$$\sin\theta = \frac{\lambda}{d}\left(m+\frac{1}{N}\right) = \frac{(\lambda+\Delta\lambda)}{d}m$$

for resolution of the spectra of order m, i.e.

$$\frac{\lambda}{\Delta\lambda} = Nm.$$

Note that the resolving power depends on the order of diffraction and the total number of lines, i.e. on the phase difference between waves from opposite ends of the grating.

2.7. Effect of source geometry

In Sections 2.4 and 2.6 we considered diffraction of a semi-parallel beam of light, such as would be produced by the arrangement shown in Fig. 2.14. There it was stated that on account of the lack of phase coherence along the length of a strip, which is a consequence of the divergence of the beam in the plane containing the strip, the diffraction pattern would be extended parallel to the source slit. The magnitude of the extension was not discussed. It can be obtained most simply by considering first the diffraction by a single slit of a plane parallel beam of light, such as is produced by a laser. In this case the diffracted intensity will be significant only in or very close to the plane containing the direction of propagation of the beam and a line in the plane of the slit normal to its length. In this plane the intensity distribution will be that of Fig. 2.7. In planes containing the length of the slit the intensity distribution will again be as shown in Fig. 2.7, but the rate of decrease of intensity with angle away from the beam direction will be the same as that in the diffraction pattern produced by an aperture of width equal to the length of the slit, and so will be very

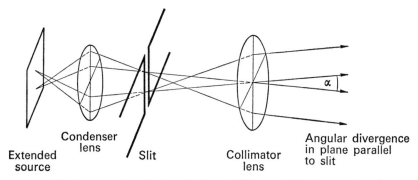

Extended source

Condenser lens

Slit

Collimator lens

Angular divergence in plane parallel to slit

FIG. 2.14. Geometry of production of a "parallel" beam of light in a spectrometer.

rapid. The diffraction pattern of a single slit produced with a plane parallel beam therefore approximates to a distribution of intensity within a very narrow ribbon.

To consider the diffraction pattern produced by a single slit when illuminated with the arrangement shown in Fig. 2.14, we suppose that each infinitesimal element of area in the collimator slit gives rise to a strictly parallel beam, and hence to a diffraction pattern of the type produced with a laser, as discussed above. If we divide the collimator slit into strips of infinitesimal width parallel to the slit, then the elements of area within each strip will give rise to diffraction patterns that are displaced relative to one another in a direction at right angles to their direction of extension, as in Fig. 2.15, and so, taken together, will build up a pattern of lines. This pattern is, of course, the geometrical image of the strip, modified by diffraction.

Other strips of the slit will produce similar patterns that are displaced laterally, as in Fig. 2.16c, and this will significantly alter the appearance of the pattern. If there is no constant phase relation between the waves from different strips then the average intensity at any point in the image plane is the sum of the intensities that would be produced by the individual strips acting alone. This is illustrated in Fig. 2.16. Figure 2.16a shows the intensity profile through the geometrical image of a slit. Each infinitesimal

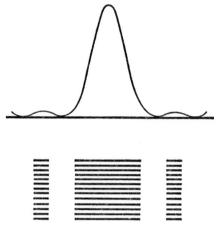

FIG. 2.15. Production of a line diffraction pattern. Elements of area at different heights in a slit source give rise to diffraction images that are displaced at right angles to their direction of extension, and so together build up a line image. The intensity profile through the line at any height is as shown above.

strip of the slit would give rise to a line image with an intensity profile shown in Fig. 2.16b, centred, of course, on the geometrical image of that strip. The broadened images from different infinitesimal strips would overlap as shown in Fig. 2.16c, leading to an intensity profile for the spectral line similar to that shown in Fig. 2.16d.

If the image of the slit is very broad in comparison with the diffraction profile of an infinitesimal strip, then the resulting image will hardly appear to be broadened at all by diffraction. If, however, the image of the slit is very much narrower than the diffraction profile, then the resulting profile will be very similar to what we will call the "theoretical" line profile. This is illustrated in Fig. 2.17.

From what has been said it is apparent that if the theoretical resolution of a diffraction grating is to be achieved in practice, then the angle sub-

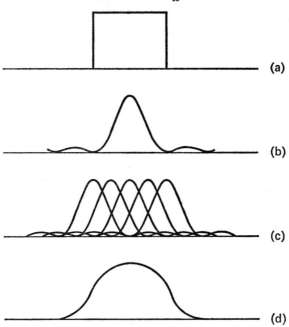

FIG. 2.16. Influence of source on observed diffraction profile. (a) Geometrical image of source. (b) Diffraction image of infinitesimal line source. (c) Superposition of images from different parts of the source. (d) Resulting intensity profile of spectral line.

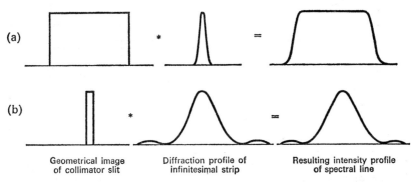

FIG. 2.17. (a) Resulting image profile with broad source and narrow diffraction profile. (b) Resulting image profile with narrow source and broad diffraction profile.

B*

tended by the width of the collimator slit at the collimator lens must be less than the angle between a principal maximum and the adjacent minimum in the diffraction pattern. For a grating of width 1 inch and for light of wavelength 5×10^{-5} cm this angle is about 2×10^{-5} radians. This means that, for a collimator lens of focal length 10 cm, the slit width must be less than 0·002 mm. The importance of the conditions of illumination in determining the practical resolution of a grating should be obvious.

2.8. Abbe theory of image formation

Figure 2.18 shows the formation of an image of a diffraction grating by an optical system. The grating is illuminated as in Fig. 2.14 with monochromatic light of wavelength λ.

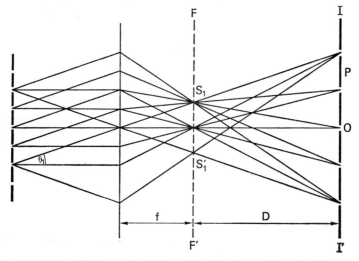

Fig. 2.18. Paths of light rays in the formation of an image of a diffraction grating by a lens.

It is apparent that all the light that goes to form the image of the grating on the screen must pass through the focal plane FF' of the lens, and furthermore that the intensity distribution in this plane is determined by the angular distribution of intensity of the light diffracted by the grating. If the grating consists of a very large number of lines the intensity will be practically zero everywhere except very close to the angles corresponding to the principal maxima. The pattern in the focal plane will therefore approximate to a series of lines whose relative intensities are determined by the angular intensity distribution function of a single slit.

If an opaque screen were to be placed in the focal plane with narrow slits in it that coincided exactly with the intensity maxima, no (or very little) light would be blocked and so the image of the grating in the image plane *II'* would not be altered. This image is therefore the same as the diffraction pattern that would be produced by the hypothetical screen when suitably illuminated.

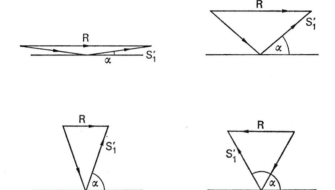

FIG. 2.19. Illustrating reversal of phase of resultant of S_1 and S_1' relative to that of S_0 ($\alpha = 0$) as *OP* (Fig. 2.18) increases.

The two first-order spectra, S_1 and S_1', will be separated by a distance $s_1 = 2f \sin \theta_1$, where f is the focal length of the lens. The amplitude due to these two sources at a point P in the image plane a distance x from O will be

$$2A_1 \cos \left(\frac{\pi s_1 x}{\lambda D} \right)$$

where A_1 is the amplitude of each of the first-order spectra and D is the distance of the image plane from the focal plane. The phase relative to the phase of a wave from S_0 will also be correctly given by this expression, i.e. 0 or π (cos = + or −), as can be seen from the sequence of amplitude–phase diagrams in Fig. 2.19. Note the abrupt reversal of phase of the resultant when the phase difference between waves from S_1 or S_2 and S_0 exceeds $\pi/2$.

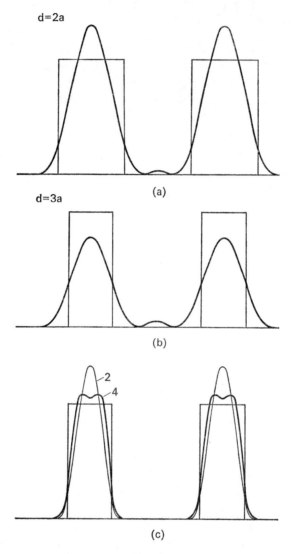

FIG. 2.20. Intensity profiles in image of diffraction grating. (a) $d = 2a$, zero and two first-order spectra only contributing to image. (b) $d = 3a$, zero and two first-order spectra only. (c) $d = 3a$, lens aperture increased to include both second-order spectra (curve 2) or both fourth-order spectra (curve 4) only.

If the aperture of the lens system is such that only the zero order and two first-order spectra are admitted, the resultant intensity in the image plane will be as shown in Fig. 2.20a, which is drawn for a slit width, a, equal to half the slit spacing, d. For this value of a/d, $A_1 = (2/\pi)A_0$, where A_0 is the amplitude of S_0. The spacing of the fringes in the image will be $d' = 2\lambda D/s_1 = (D/f)d$, i.e. the magnification is D/f, which is just the magnification to be expected from elementary lens theory.

The way in which the image changes as the lens aperture is increased to include more spectra is shown in Fig. 2.20 b and c, which are drawn for the case $d = 3a$. Each successive pair of spectra contribute a sinusoidally varying amplitude in the image plane with a repeat distance that is inversely proportional to the separation of the spectra in the focal plane, i.e. to the order of the spectra.

Two important conclusions can be drawn from this treatment; detail on a finer scale than the repeat distance of the amplitude variation from the highest-order spectra passed by the lens will not appear in the image, and false detail, i.e. detail not present in the object, may appear. The smallest detail discernible in the object will be on a scale

$$\delta = \frac{2\lambda D}{s_m}\frac{f}{D},$$

since the magnification is D/f, i.e. $\delta = \lambda/\sin\theta_m$, where θ_m is the angle of the highest-order spectra passed. This, then, is the resolving power of the lens.

2.9. The sinusoidal grating

A grating of very considerable importance in the theory of diffraction is one in which the amplitude of the scattered wavelets varies sinusoidally along the length of the grating, the amplitude of the wavelet scattered by a strip of width dy at position y being

$$A_{m,y}\,dy = A_m \sin(2\pi my/d)\,dy.$$

The scattered wavelets are either in phase or exactly out of phase with each other, depending on the relative signs of $A_{m,y}$. When such a grating is illuminated at normal incidence with a plane wave of wave number k, the complex amplitude of the wave scattered at an angle θ will be

$$A_\theta = \int A_m \sin\left(2\pi my/d\right) \exp\left(2\pi iky \sin\theta\right) dy$$

$$= \int A_m \sin\left(2\pi my/d\right) \cos\left(2\pi ky \sin\theta\right) dy$$

$$+ i \int A_m \sin\left(2\pi my/d\right) \sin\left(2\pi ky \sin\theta\right) dy$$

where each integral is taken over the extent of the grating.

Now

$$\sin A \cdot \cos B = \tfrac{1}{2}\left[\sin\left(A+B\right) + \sin\left(A-B\right)\right]$$

and

$$\sin A \cdot \sin B = \tfrac{1}{2}\left[\cos\left(A+B\right) - \cos\left(A-B\right)\right].$$

$$\therefore \quad A_\theta = \tfrac{1}{2}A_m\left[\int \sin\left(2\pi y(m/d + k\sin\theta)\right) dy + \int \sin\left(2\pi y(m/d - k\sin\theta)\right) dy\right]$$

$$+ \tfrac{1}{2}iA_m\left[\int \cos\left(2\pi y(m/d + k\sin\theta)\right) dy\right.$$

$$\left. - \int \cos\left(2\pi y(m/d - k\sin\theta)\right) dy\right].$$

Consider now one of these integrals, $\int \cos\left(2\pi y(m/d - k\sin\theta)\right) dy$. If the integral is taken over a distance that is an exact multiple of $(m/d - k\sin\theta)^{-1}$, then the mean value of the integrand, and hence the value of the integral, will be zero. This will not be possible, of course, if $k\sin\theta = m/d$, when the integrand is not periodic but has the value unity. If $k\sin\theta \neq m/d$ the whole grating can be divided into two parts, one of which is an exact multiple of $(m/d - k\sin\theta)^{-1}$ and the other is the remainder of the grating, less than $(m/d - k\sin\theta)^{-1}$. The integral over the whole grating will then be equal to the integral over the remainder section, the first part contributing nothing. If the grating is large in comparison with $(m/d - k\sin\theta)^{-1}$, then the integral over the remainder section will be small in comparison with the integral over the whole grating for an angle θ such that $m/d = k\sin\theta$. In the limit of an infinite grating the integral will be infinitesimal in comparison with its value at $\theta = \sin^{-1}(m/kd)$ for all other angles.

The other integrals can be treated similarly. The first two will never be appreciable, whilst the third will be large at an angle such that

$$k\sin\theta = -m/d.$$

The incident beam will therefore be split by the grating into two beams, making angles of $\theta = \pm\sin^{-1}(m/kd)$ with the incident beam direction. At these angles $A_\theta = \tfrac{1}{2}A_m$. At other angles the intensity of diffraction will be negligible. This is illustrated in Fig. 2.21.

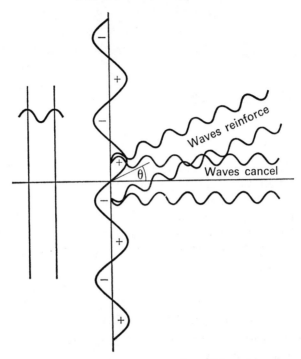

Fig. 2.21. Diffraction by a sinusoidal grating.

CHAPTER 3

TWO-DIMENSIONAL DIFFRACTION

3.1. Introduction

In this chapter we will consider some examples of diffraction by apertures and arrays of scattering centres that are two-dimensional, and derive some results that will be of use when considering diffraction by crystals. First, however, we will consider the propagation of a spherical wave and show that application of Huygens' principle leads to results that are in agreement with the known facts provided that we make certain assumptions concerning the amplitude and phase of the secondary wavelets.

3.2. Propagation of a spherical wave

Consider a spherical wave originating at a point O (Fig. 3.1). The displacement at a point P will be given by

$$\phi_P = \frac{a}{r} \exp \left[2\pi i(kr - vt)\right]. \tag{3.1}$$

The displacement can also be expressed as the sum of the displacements due to the secondary wavelets originating on some such wavefront as that shown at a distance r_0. The contribution to the complex amplitude at P from an element of the wavefront of area dS at the position shown will be

$$dA = \frac{a'f(\theta)\, dS}{r'} \exp \left[2\pi ik(r' + r_0 - r)\right]$$

where the phase of the wavelet originating from O' has been taken as zero, and a' and $f(\theta)$ are to be determined. We may take for dS the annular element of width $r_0\, d\psi$ at an angle ψ to OP: $dS = 2\pi r_0^2 \sin \psi\, d\psi$. Then

$$A = \int_0^\pi \frac{2\pi r_0^2 a'f(\theta) \sin \psi}{r'} \exp \left[2\pi ik(r' + r_0 - r)\right] d\psi.$$

The integral can be evaluated most easily from a consideration of the
amplitude–phase diagram, which is its geometrical representation (Fig. 3.2).

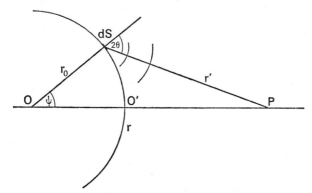

Fig. 3.1. Propagation of a spherical wave.

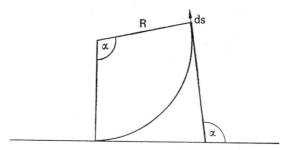

Fig. 3.2. Part of amplitude–phase diagram for situation of Fig. 3.1.

The element dS will contribute a segment of length

$$ds = \frac{2\pi r_0^2 a' f(\theta) \sin \psi \, d\psi}{r'}$$

at an angle $\alpha = 2\pi k(r' + r_0 - r)$. The radius of curvature at this part of
the curve is therefore

$$R = \frac{ds}{d\alpha} = \frac{r_0^2 a' f(\theta) \sin \psi \, d\psi}{r' k \, dr'} \; .$$

Now

$$r'^2 = r^2 + r_0^2 - 2rr_0 \cos \psi.$$

$$\therefore \quad 2r' dr' = 2rr_0 \sin \psi \, d\psi$$

and so

$$R = \frac{r_0}{r} \frac{a'}{k} f(\theta).$$

If $f(\theta)$ is a slowly varying function of θ (slow, that is, in comparison with α), R will be approximately constant while α increases by 2π and so the amplitude–phase diagram will approximate to a circle. The radius of this circle will, however, get progressively smaller, as shown in Fig. 3.3, on account of the obliquity factor, $f(\theta)$.

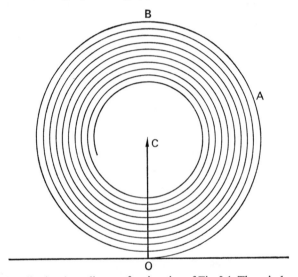

FIG. 3.3. Amplitude–phase diagram for situation of Fig. 3.1. The spiral eventually converges to the point C, giving the resultant shown. The rate of convergence is neither as rapid nor as uniform as indicated.

For the integral to converge the end of the curve must approach a point, i.e. $R \to 0$, and so $f(\theta) \to 0$ as $2\theta \to \pi$. This condition is satisfied by the obliquity factor $f(\theta) = \cos^2 \theta$ of the Kirchhoff integral. The asymptotic point is the centre of the circle. The resultant amplitude at P is therefore the initial radius of the spiral. Equating this to the amplitude given by eq. (3.1) we find

$$\frac{a}{r} = \frac{r_0}{r} \frac{|a'|}{k} f(0)$$

and so

$$| a' | = \frac{a}{r_0} \cdot \frac{1}{\lambda}.$$

We see, then, that the amplitude of a secondary wavelet is proportional to the amplitude of the incident wave, a/r_0, and inversely proportional to the wavelength, as already stated. Also, the phase of the resultant is $\pi/2$ behind that of the wavelet from O'. This wavelet travels the same distance as the original wave, however, between O' and P, and so the phase of a secondary wavelet must be $\pi/2$ ahead of the incident wave.

3.3. Fresnel zones

Consider the wavefront of Fig. 3.1 divided into zones such that the phases at P relative to the phase of the wavelet from O' of all wavelets from elements of area in the first zone lie between O and π, in the second

FIG. 3.4. Contributions to resultant amplitude at P (Fig. 3.1) from successive half-period zones (schematic). The vectors have been displaced for clarity.

zone from π to 2π, and so on. The boundary between the first and second zones will be at a distance $r - r_0 + \lambda/2$ from P, that between the second and third at $r - r_0 + 2\lambda/2$, etc., so the boundaries of the zones are concentric circles. The first zone will contribute the portion OAB of the amplitude–phase diagram, with resultant OB. The resultants of successive zones are as shown in Fig. 3.4, where they have been displaced for clarity. Such zones are known as Fresnel or half-period zones.

Had all the wavelets from elements of area in the first zone arrived at P in phase the amplitude–phase diagram would have been a straight line of length equal to the arc OAB. This arc is of length πR, and so we see that the resultant amplitude due to the whole wavefront is just $1/\pi$ times the amplitude that would have resulted from the first zone alone had all the wavelets from the first zone been in phase.

The resultant of the nth zone is just $2R_n$, where R_n is the radius of the corresponding part of the amplitude–phase diagram. The decrease in R_n with increasing n is due solely to the obliquity factor, $f(\theta)$, so we see that the area of a zone is proportional to its distance from P, these two factors in the amplitude exactly cancelling. If $r' \gg \lambda/2$, the areas of successive zones will be very nearly the same, as may be shown directly. The radius of the outer boundary of the nth zone is therefore proportional to $n^{\frac{1}{2}}$.

A diffracting screen constructed in such a way that it obstructs alternate zones between concentric rings whose radii increase as $n^{\frac{1}{2}}$ is known as a zone plate. Consider a plane wave incident normally on such a screen. For a point P on the axis at some distance (determined by the wavelength of the light) from the screen the zones of the screen will coincide with the half-period zones of a wavefront at the position of the screen. As alternate zones will be obscured the amplitude at P will be very large, the action of the zone plate being to focus the light. There will be other foci at distances such that each zone of the plate passes 3 or 5 or 7, etc., half-period zones.

3.4. Scattering by a plane of atoms

3.4a. *Forward scattering*

Consider a spherical wave $\phi = (a_0/r) \exp[2\pi i(kr - vt)]$ from a source S incident on an infinite plane of atoms. We wish to calculate the amplitude of the scattered wave at a point P such that SP is normal to the plane of atoms (Fig. 3.5).

Let there be n atoms per unit area in the plane. Each will give rise to a scattered wave whose amplitude at a distance r' will be given by $\phi = \phi_0 af(\theta)/r'$, where ϕ_0 is the amplitude of the incident wave and $af(\theta)$ is the scattering power of the atom, which will depend on the nature of the incident radiation. The annular element of area between circles of radii y and $(y+dy)$ in Fig. 3.5 will therefore contribute a complex amplitude dA at P such that

$$dA = ds\, e^{i\alpha}$$

where

$$ds = \frac{a_0}{r'_1} \frac{af(\theta)}{r'_2}\, n2\pi y\, dy$$

and

$$\alpha = 2\pi k(r'_1 + r'_2 - r_1 - r_2).$$

The origin of phase is chosen such that $\alpha = 0$ for the scattered wave from O.

The radius of curvature of the corresponding part of the amplitude–phase diagram is

$$R = \frac{ds}{d\alpha} = \frac{a_0 a f(\theta) n y \, dy}{r'_1 r'_2 k(dr'_1 + dr'_2)}.$$

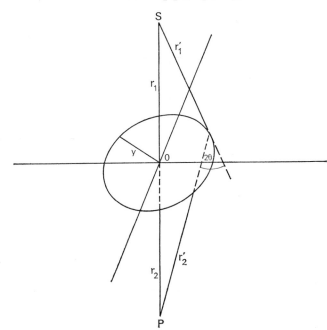

FIG. 3.5. Geometry of scattering by a plane of atoms at normal incidence.

Now

$$y^2 = r'^2_1 - r^2_1 = r'^2_2 - r^2_2,$$

$$2y \, dy = 2r'_1 \, dr'_1 = 2r'_2 \, dr'_2,$$

$$R = \frac{n a_0 a f(\theta)}{k(r'_1 + r'_2)}.$$

The amplitude–phase diagram therefore approximates to a circle, as before. The question of convergence will be considered in Section 3.4d. We shall assume for the time being that the amplitude–phase diagram does converge, to the centre of the initial circle, so that the amplitude of the scattered wave at P will be equal to the initial radius of curvature, i.e.

$$A = a_0 \frac{n\lambda a f(0)}{(r_1 + r_2)}.$$

For an incident plane wave, for which $r_1 \gg r_2$, this becomes

$$A = \phi_0 n\lambda a f(0).$$

If SP makes an angle θ with the plane of atoms, as in Fig. 3.6, the boundaries of the first few half-period zones become ellipses, the major and minor semi-axes of the ellipse bounding the mth zone being given by

$$a_m \sin \theta = \left(\frac{m\lambda r_1 r_2}{r_1 + r_2} \right)^{\frac{1}{2}} = b_m,$$

as can easily be shown from Fig. 3.6b. This is valid provided that $a_m, b_m \ll r_1, r_2$.

The area of the first half-period zone is therefore

$$\pi a_1 b_1 = \frac{\pi\lambda}{\sin\theta} \frac{r_1 r_2}{r_1 + r_2}$$

and the amplitude of the scattered wave at P is

$$A = \frac{1}{\pi} \frac{a_0}{r_1} \frac{a f(0)}{r_2} n \frac{\pi\lambda r_1 r_2}{\sin\theta(r_1 + r_2)}.$$

For an incident plane wave this becomes

$$A = \phi_0 \frac{n\lambda a f(0)}{\sin\theta}. \tag{3.2}$$

The phase of the resultant at P is $\pi/2$ behind that of the scattered wave from O, as in Section 3.3. We can therefore write

$$A = \phi_0 q_0 \, e^{i\pi/2}$$
$$= \phi_0 i q_0$$

where

$$q_0 = \frac{n\lambda a f(0)}{\sin\theta}.$$

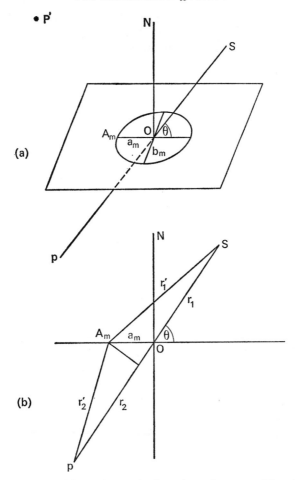

FIG. 3.6. Geometry of forward scattering by a plane of atoms at oblique incidence. OS, ON and OA_m are coplanar.

3.4b. *Reflection*

In addition to a scattered wave in the direction of the incident wave there will be a reflected wave. At the point P' in Fig. 3.6, which is the mirror image of P in the atom plane, for any point O' in the plane the distances $O'P$ and $O'P'$ will be equal. Therefore the phases at P and P' of the wave scattered from O' will be the same, and the geometry of the half-period zones will be identical with the case just considered. Since for all atoms in the first half-period zone the angle of scattering is very close to 2θ, the amplitude of the scattered wave at P' will be

$$A_{P'} = \phi_0 \frac{n\lambda a f(\theta)}{\sin \theta}.$$

The phase of the resultant is again $\pi/2$ behind that of the scattered wave from O, so we can write

$$A_{P'} = \phi_0 i q$$

where

$$q = \frac{n\lambda a f(\theta)}{\sin \theta}.$$

3.4c. *Other scattered beams*

If the atoms were continuously distributed on the plane, or if their separation were less than half the wavelength of the incident radiation, then the forward scattered and reflected waves would be the only scattered waves of significant amplitude. Atoms in crystals, however, are not continuously distributed but are arranged in a regular array, with separations that are generally greater than the wavelengths of incident radiations other than light, and so there may be other scattered waves.

Suppose that all the atoms are located on lines of spacing d drawn parallel to the minor axes of the half-period zones of the forward scattered and reflected waves, as in Fig. 3.7. Then at a point P'' distant r_2 from O and in the plane containing the points S and O and the normal to the plane at O, such that OP'' makes an angle θ'' with the plane, the phase difference between waves from O and O' will be

$$\alpha = 2\pi k(r_1' + r_2'' - r_1 - r_2)$$

$$= 2\pi k \left\{ y(\cos \theta + \cos \theta'') + \frac{y^2}{2} \left(\frac{\sin^2 \theta}{r_1} + \frac{\sin^2 \theta''}{r_2} \right) \right\}.$$

Now at the positions of the atoms we can put $y = ld$, where l is an integer, and so the first term on the right becomes $2\pi k \cdot ld(\cos \theta + \cos \theta'')$. If now $d(\cos \theta + \cos \theta'') = p\lambda$, where p is an integer, this term becomes a multiple of 2π and so can be ignored. The second term on the right, varying as y^2, is identical with the corresponding phase difference for the forward scattered and reflected waves if $\sin^2 \theta = \sin^2 \theta''$, and is never very different. The wave scattered in the direction θ'' will therefore be comparable in intensity with the other scattered waves, any difference arising mainly through the different value of $f(\theta)$.

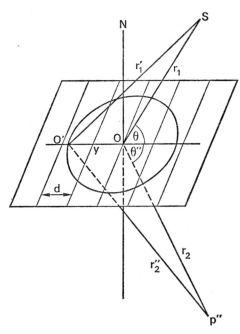

FIG. 3.7. Geometry of scattering by a plane of atoms in a crystal. Atoms are situated only on the lines shown.

3.4d. *Validity of Fresnel zone treatment*

The assumption that the resultant of the scattered waves from the whole plane will be half that from the first Fresnel zone, with a phase shift of $\pi/2$ relative to the wave from the centre of the zone, will be valid only if the amplitude–phase diagram converges to the centre of the initial circle. Convergence, however, is very slow in the cases considered. For the case of normal incidence the radius of curvature of the amplitude–phase diagram will decrease on account of the factor $f(\theta)/(r_1' + r_2')$, but this does not go to zero until r_1' and r_2' become infinite: $f(\theta)$ remains finite at all angles, and indeed may not decrease significantly at all. For oblique incidence the

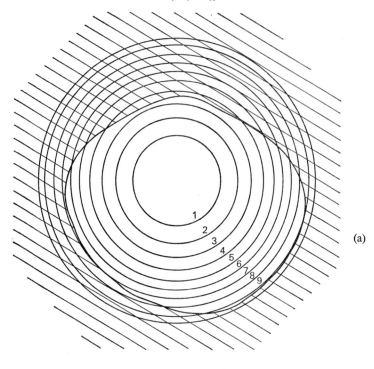

(a)

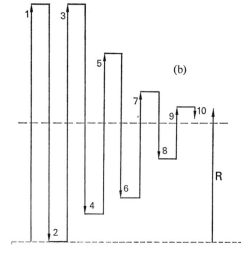

(b)

Fig. 3.8. (a) An irregular aperture
in an opaque screen, with the
boundaries of the Fresnel zones
shown. (b) The corresponding
amplitude–phase diagram. A more
accurate analysis, for instance in
quarter-period zones, would show
that the resultants of partial zones
are not strictly proportional to the
uncovered areas of the zones, nor
are their phases exactly as shown.

problem is more complicated, since the boundaries of the zones are not elliptical, even approximately, if $a_m \sim r_1$, and also the distance and obliquity factors will vary substantially within a zone. It is clear, though, that convergence will be no more rapid on account of this.

What the treatment predicts is that the intensity at P will depend on the relation between the boundaries of the atom plane and the Fresnel zones. In general the boundary will not lie wholly within one zone, nor even within a small number of zones. Since, under these circumstances, the resultant amplitude from a zone will be proportional to the fraction of the zone that is covered by atoms, there will be a steady decrease in the amplitudes of successive zones when the zones begin to overlap the boundary of the atom plane. If the decrease is spread over a large number of zones it is easy to see that the amplitude–phase diagram will converge to a point very close to the centre of the initial circle. The exact point will depend on the details of the boundary. This is illustrated in Fig. 3.8.

If, by chance, the boundary of the atom plane happens to lie within only a very few zones (in an extreme case, within one zone), then there will be variations in intensity around P. In practice this would not be observed, even with measuring instruments of unlimited resolution, since the angle between the directions of maximum intensity would be very much less than the angle subtended by the source, unless the atom plane were very small indeed.

Another way of looking at the problem is that in the radiation from an extended source phase coherence is not maintained over long distances, because neither are the wavefronts strictly planar, nor is their form constant with time. The reason for this is illustrated in Fig. 3.9. Waves from opposite ends of the source reach points A and B with different phases, and so the phase of the disturbance at B relative to that at A will depend on the relative phases of waves from different parts of the source. The effect of this is that the•individual wavelets from elements of area between circles of radii y and $(y+dy)$ (Fig. 3.5) will not all be exactly in phase at P, the scatter in phase increasing with y, and so their resultant will be less than that calculated by the simple Fresnel theory. The contribution from successive Fresnel zones is therefore also reduced and the amplitude–phase diagram converges rapidly.

We can obtain a simple estimate of the distance over which phase coherence is maintained if we suppose that phases are completely coherent for a distance Y but beyond this are completely incoherent. The angular width of the diffraction maximum from an aperture of width Y will then be equal to the divergence of the incident beam. For an incident X-ray beam of wavelength 1·5 Å and divergence 10^{-3} radian this is $Y \simeq 1500$ Å: for an electron beam of wavelength 0·037 Å (100 kV) and the same divergence $Y \simeq 40$ Å.

The assumption that the amplitude of the wave scattered from an element of area is proportional to its area also needs some justification.

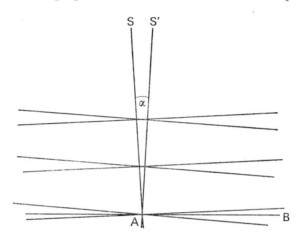

FIG. 3.9. Waves from S and S', parts of an extended source subtending an angle α at A, are in phase at A but differ in phase at B. At a later time the relative phases of S and S', which are independent sources, may have changed. Therefore neither the amplitude nor the phase of the disturbance along the line AB, a "wavefront", will be strictly constant.

It will be valid provided that the number of atoms in an area over which the phase of the scattered waves does not change much is large. Since the area of a Fresnel zone is $\sim \pi \lambda r_1 r_2 /(r_1 + r_2)$, and the number of atoms per square centimetre of an atom plane is $\sim 10^{15}$, this assumption is seen to be quite justified; for $r_1 = r_2 = 10$ cm and $\lambda = 1·5$ Å, the number of atoms in a zone is about 2×10^8.

3.5. Diffraction by an aperture

In this section we will consider diffraction of a plane parallel beam of light under Fraunhofer conditions.

In Fig. 3.10 a plane parallel beam of light is incident normally on a screen in which there is an aperture: we wish to determine the intensity in some arbitrary direction specified by the angles θ and ψ.

A general method of doing this is as follows: take Cartesian axes x, y and z with the x-axis parallel to the incident beam direction and the y-axis in the plane defined by the x-axis and the direction in which we wish to determine the intensity. The phase difference between wavelets from

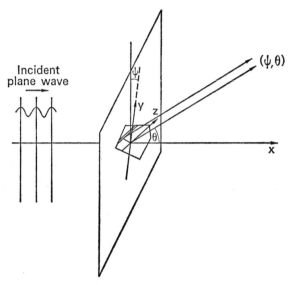

Fig. 3.10. Geometry of diffraction of a plane wave by an aperture.

elements of area at the origin and at the point (y, z) in the aperture will then be $\alpha = 2\pi k y \sin \theta$. The resultant amplitude from the whole aperture will then be

$$A = \text{const.} \iint \exp (2\pi i k y \sin \theta)\, dy\, dz$$

$$= \text{const.} \int \left(\int dz\right) \exp (2\pi i k y \sin \theta)\, dy$$

$$= \text{const.} \int l(y) \exp (2\pi i k y \sin \theta)\, dy$$

where $l(y)$ is the length of the strip of aperture parallel to the z-axis lying between the limits y and $(y+dy)$ and is a function of y. The integral is taken over the extent of the aperture.

To illustrate the method we will consider the variation of diffracted intensity with angle for a circular aperture of radius r. If we take the origin at the centre of the aperture, $l = 2(r^2 - y^2)^{\frac{1}{2}}$, and so

$$A = \text{const. } \int_{-r}^{r} (r^2 - y^2)^{\frac{1}{2}} \exp(2\pi i k y \sin \theta) \, dy$$

$$= \text{const. } \int_{-r}^{r} (r^2 - y^2)^{\frac{1}{2}} \cos(2\pi k y \sin \theta) \, dy$$

$$+ i \int_{-r}^{r} (r^2 - y^2)^{\frac{1}{2}} \sin(2\pi k y \sin \theta) \, dy.$$

Since the integrand of the second integral is an odd function of y, the value of the integral between the limits $\pm r$ is zero. The integrand of the first integral is an even function of y, and so

$$A = 2 \times \text{const. } \int_{0}^{r} (r^2 - y^2)^{\frac{1}{2}} \cos(2\pi k y \sin \theta) \, dy.$$

The obliquity factor, which is independent of y, is included in the constant in the above expression. It is, of course, not independent of θ. However, in most problems of interest the range of angles involved is very small and so the variation of the obliquity factor can be neglected. If we put $A = A_0$ at $\theta = 0$, the expression for A becomes

$$A = \frac{4A_0}{r^2} \int_{0}^{r} (r^2 - y^2)^{\frac{1}{2}} \cos(2\pi k y \sin \theta) \, dy.$$

The integral can easily be evaluated numerically for different values of $k \sin \theta$. The resulting intensity distribution is, of course, the same as the image of a star formed by a telescope, since the star itself subtends an angle that is very small in comparison with the angular width of the diffraction pattern formed by the telescope objective. It was first derived by Airy, in 1834; an alternative treatment is given by Longhurst (1957).

The intensity distribution in any plane containing the incident beam direction is similar to, though not identical with, the single slit diffraction pattern. The first minimum occurs at an angle θ such that $\sin \theta = 0.61\lambda/r = 1.22\lambda/d$, where d is the diameter of the aperture.

With an aperture of high symmetry it may be easier to use a method that utilises this symmetry. For instance, for a rectangular aperture of size $\mathbf{a} \times \mathbf{b}$, if we divide the aperture into strips parallel to, say, the side of length $\mathbf{a}$, then in any direction making an angle $(\pi/2 - \theta_1)$ with the length of the strips, the amplitude of the resultant wavelet from each strip will be

$$dA = \text{const. } \frac{\sin(\pi a \sin \theta_1/\lambda)}{\pi a \sin \theta_1/\lambda},$$

each strip acting, for diffraction in a plane containing its length, as a single slit of width equal to its length. The phase of the resultant will be the phase of the centre element. In a direction making also an angle $(\pi/2 - \theta_2)$ with the side of length **b** the resultant amplitude due to the whole aperture will be

$$A = \text{const.} \frac{\sin(\pi a \sin\theta_1/\lambda)\sin(\pi b \sin\theta_2/\lambda)}{(\pi a \sin\theta_1/\lambda)(\pi b \sin\theta_2/\lambda)} .$$

3.6. Diffraction by arrays of apertures

3.6a. *Irregular arrays*

If there is no regularity in the arrangement of a large number of identical apertures in a plane screen, then in any direction the phases of the waves from the different apertures will be random and so the intensity of the wave will be equal to the intensity produced by any one aperture multiplied by the number of apertures. This method of increasing the intensity of a single aperture diffraction pattern can be useful if it is desired to record the pattern photographically.

3.6b. *Regular arrays*

Consider first an array of point apertures. The points can be thought of as lying on two non-parallel sets of lines. Intense diffracted beams will be produced if the waves from all the apertures are in phase and this condition will be satisfied if the waves from the apertures on any two lines, one from each set, are in phase. We have already considered the case of a set of scattering centres uniformly spaced on a straight line: intense diffracted beams are produced at angles θ such that $d \sin\theta = m\lambda$, where d is the spacing of the scattering centres along the line and m is any integer. This condition will be satisfied for all directions on the surface of a cone of semi-angle $\pi/2 - \theta$, with the line as axis. For a two-dimensional array of scattering centres, then, strong diffracted beams will be produced in directions that are the intersections of the two sets of cones, one set for each of the two lines. Away from these directions the intensity will decrease, the amplitude being the product of two factors of the form of eq. (2.1) if the array is a parallelogram with sides parallel to the two sets of lines.

An array of finite apertures is equivalent to an array of point sources in which the wave from each source has the angular distribution of amplitude appropriate to the aperture. Diffraction maxima occur at the same angles as before, the intensities being proportional to the intensity of the wave scattered in those directions by a single aperture. The sharpness of the diffraction maxima will depend on the number of units in the array.

3.6c. *Application to structure analysis*

It can be shown that if a crystal be rotated in an X-ray beam about an axis at right angles to the beam, then the intensities of the diffracted beams in the plane normal to the axis of rotation depend only on the normal projection onto this plane of the electron density distribution in the crystal; the distribution along lines parallel to the axis of rotation has no effect on the intensities of these beams, since electrons on any one line will always scatter in phase. Now if a screen be prepared which is a photographic enlargement of this projection, then the diffraction pattern produced by the screen when suitably illuminated will consist of a number of beams that correspond exactly with the beams diffracted by the crystal. It is, of course, never possible to produce a screen that matches exactly the projection of the electron density. Instead, this must be represented by scattering centres that simulate the atomic scattering factors, placed at the projected positions of the atom centres. The simplest approach, to use large holes to represent heavy atoms and small ones for light atoms, is unsatisfactory since the amplitude of the diffracted wave from a large aperture falls off more rapidly with angle than that from a small one, which does not simulate the atomic scattering factors. This difficulty can be overcome to some extent by representing the atoms by rings, adjusting both the diameter and thickness of the rings to match the magnitude and rate of decrease of the atomic scattering factors. Other refinements have also been employed.

This provides a very quick and simple way of checking the correctness of a postulated structure and, with some experience, of suggesting modifications if it is found not to be correct.

The number of units in any crystal is vastly greater than could be incorporated into a screen. Fortunately, for a comparison of intensities, it is not necessary to have anything like this number of units in a screen. Furthermore, in practice the sharpness of the diffraction maxima from a screen will depend on the conditions of illumination as well as on the number of units in the screen; there is no point in trying to reduce the angular widths of the maxima to less than the divergence of the illuminating beam. In fact, a useful comparison with the X-ray intensities can be made with diffracted beams of much greater angular widths, requiring diffracting screens with far fewer units.

It is worth noting that projections of electron density cannot be obtained in the same way from measurements of the intensities of diffracted beams of neutrons or electrons. It is possible with X-rays because the coherent scattering of X-rays, which alone contributes to the diffraction peaks, is by electrons alone. This is not true of the scattering of neutrons or electrons.

CHAPTER 4

THREE-DIMENSIONAL DIFFRACTION

4.1. Introduction

Diffraction by crystals differs in two important respects from diffraction by one- or two-dimensional assemblies: firstly, unless the crystal is suitably oriented with respect to the incident beam there will be no significant diffracted beams at all, and secondly, a scattered wavelet from part of a crystal may suffer further scattering before emerging from the crystal. The importance of rescattering, in particular of scattering back into the incident beam direction, will depend on the relative intensities of the incident and diffracted beams. This will vary from place to place within the crystal, but if at all places the intensity of the diffracted beam is small in comparison with the intensity of the incident beam, then re-scattering may be neglected, as may also the loss of intensity from the incident beam. This approximation gives rise to what is known as the kinematical theory of diffraction. The more exact treatment in which rescattering is taken into account, known as the dynamical theory, is unfortunately much more difficult, both as regards derivation and application.

The treatment of this chapter will assume a knowledge of elementary crystallography and the reciprocal lattice. Those not familiar with these topics will find a brief account in Chapter 7.

4.2. Geometry of diffraction

Consider a plane wave of wave vector **k** incident on a crystal in which there is one atom per lattice point. Intense diffracted beams will occur in directions such that all the scattered waves are in phase. In Fig. 4.1 the phase difference between waves scattered by atoms at the origin and at the point A is

$$\alpha = \frac{2\pi}{\lambda}(OC - AB)$$

$$= \frac{2\pi}{\lambda} \left(\frac{\mathbf{r} \cdot \mathbf{k}'}{|\mathbf{k}'|} - \frac{\mathbf{r} \cdot \mathbf{k}}{|\mathbf{k}|} \right)$$

$$= 2\pi \mathbf{r} \cdot (\mathbf{k}' - \mathbf{k})$$

since $|\mathbf{k}'| = |\mathbf{k}| = 1/\lambda$.

The waves will be in phase if $\mathbf{r} \cdot (\mathbf{k}' - \mathbf{k})$ is an integer. Since A is a lattice point we can write

$$\mathbf{r} = n_1 \mathbf{a}_1 + n_2 \mathbf{a}_2 + n_3 \mathbf{a}_3$$

where n_1, n_2 and n_3 are integers and $\mathbf{a}_1$, $\mathbf{a}_2$ and $\mathbf{a}_3$ are the base vectors

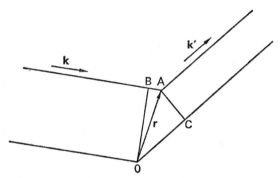

Fig. 4.1. Geometry of phase difference between waves scattered by two atoms.

defining the crystal lattice. If we refer $(\mathbf{k}' - \mathbf{k})$ to $\mathbf{b}_1$, $\mathbf{b}_2$ and $\mathbf{b}_3$, the base vectors defining the reciprocal lattice, then

$$\mathbf{k}' - \mathbf{k} = \mathbf{K} = K_1 \mathbf{b}_1 + K_2 \mathbf{b}_2 + K_3 \mathbf{b}_3$$

and

$$\mathbf{r} \cdot \mathbf{K} = n_1 K_1 + n_2 K_2 + n_3 K_3.$$

This will be integral for all integral values of n_1, n_2 and n_3 if, and only if, K_1, K_2 and K_3 are also integral, which implies that $\mathbf{K}$ is a vector from the origin to one of the reciprocal lattice points.

It is possible to give a simple geometrical interpretation of the condition for the existence of a strong diffracted beam, as follows: draw a vector $-\mathbf{k}$ from any reciprocal lattice point, O, as origin and with the end of this vector, O', as centre draw a sphere of radius $|\mathbf{k}|$ (Fig. 4.2). If this sphere passes through a reciprocal lattice point, say the point $G(h, k, l)$, then there will be an intense diffracted beam with wave vector $\mathbf{k}' = \overrightarrow{O'G}$.

The sphere may, of course, pass through more than one reciprocal lattice point, in which case there will be more than one strong diffracted beam, or it may pass through none at all, when there will be none. This construction is due originally to Ewald and is known as the Ewald or reflecting sphere construction.

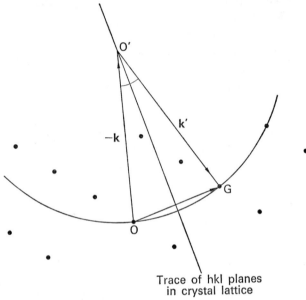

FIG. 4.2. The reflecting sphere construction in the reciprocal lattice.

Now the vector $\mathbf{g}$ from the origin to the reciprocal lattice point G is normal to the set of crystal lattice planes with Miller indices hkl and is of length $|\,g\,| = 1/d_{hkl}$, where d_{hkl} is the spacing of the crystal lattice planes. From this it follows (see Fig. 4.2) that the incident and diffracted beams make equal angles, θ, with the crystal lattice planes, and that

$$|\,\mathbf{g}\,| = 2\,|\,\mathbf{k}\,|\,\sin\,\theta,$$

i.e. $\lambda = 2d_{hkl}\sin\,\theta.$ (4.1)

The diffracted beam can be thought of as a reflection from the set of crystal lattice planes with indices hkl. Obviously, all atoms within one plane of the set will scatter in phase if the angles of incidence and reflection are the same (as here). Equation (4.1) states that a strong diffracted beam will be produced when the path difference between waves scattered from atoms in adjacent planes is equal to λ, as can be seen from Fig. 4.3. Equation (4.1) is the Bragg law: the glancing angle of incidence, θ, is known as the Bragg angle for the particular wavelength and set of planes involved.

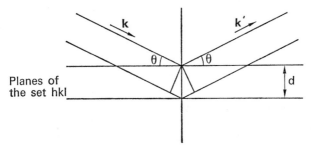

Fig. 4.3. Geometry of strong diffracted beam. Waves reflected from successive atom planes will reinforce if the path difference between them $\equiv 2d \sin \theta = \lambda$.

The treatment given in this section emphasises that the phenomenon is one of diffraction rather than of reflection, but it also shows that there will always be a set of planes bisecting the angle between the incident beam and any strong diffracted beam, which can therefore always be looked upon as a reflection from this set of planes. Following the usual custom, we shall speak of the diffracted beam as having been reflected by the *hkl* set of planes.

If there is more than one atom associated with each lattice point, as is usually the case, then the amplitude and phase of the resultant wave from the group associated with the lattice point will vary in a complex manner with the angle of diffraction. They will, however, be the same for every group, since the groups themselves are identical, and so the directions of the diffracted beams from the whole crystal will be the same as for a crystal with only one atom per lattice point.

4.3. Kinematical theory of intensity of diffraction

4.3a. *Introduction*

Any measurement of the intensity of a diffracted beam will involve the intensity of diffraction at angles other than the exact Bragg angle, if only because there will always be a finite divergence in any incident beam. It is therefore important to know how the intensity of diffraction varies with angle. We shall now derive an expression for the intensity of a diffracted beam when the Bragg equation is nearly but not exactly satisfied, and use it to calculate the total energy reflected when a small crystal is rotated in a beam of radiation through a sufficient angle to cover all angles at which appreciable diffraction takes place. This energy is commonly referred to as the integrated intensity, despite the fact that it is not, strictly speaking, an intensity at all. A better term would, perhaps, be *integrated reflection*, but the term *integrated intensity* is too well established to be changed now, and will be used hereinafter.

Basically the same quantity is measured if a stationary crystal is bathed in a beam of radiation of sufficient divergence to cover all angles of incidence at which a diffracted beam of appreciable intensity is produced, provided that the intensity per unit solid angle of the incident beam is constant over this range. In this case what is measured is, strictly speaking, the power of the diffracted beam.

4.3b. *General formulation*

Consider a beam of radiation of wave vector $\mathbf{k}$ incident on a crystal. We will suppose that the lattice points of a primitive lattice are at the positions defined by the vectors $\mathbf{r}_1, \mathbf{r}_2, \ldots, \mathbf{r}_j, \ldots, \mathbf{r}_N$ from some fixed origin in the crystal and that in each unit cell the atoms are at positions defined by the vectors $\boldsymbol{\rho}_1, \boldsymbol{\rho}_2, \ldots, \boldsymbol{\rho}_p$ from the origin of that unit cell. The atomic positions relative to the origin of the lattice are then given by the vectors $\mathbf{r}_1 + \boldsymbol{\rho}_1$, $\mathbf{r}_1 + \boldsymbol{\rho}_2$, $\ldots$, $\mathbf{r}_1 + \boldsymbol{\rho}_p$, $\ldots$, $\mathbf{r}_j + \boldsymbol{\rho}_1$, $\mathbf{r}_j + \boldsymbol{\rho}_2$, $\ldots$, $\mathbf{r}_j + \boldsymbol{\rho}_p$, $\ldots$, $\mathbf{r}_N + \boldsymbol{\rho}_p$. In order to calculate the intensity of the diffracted beam at some remote point P we will make the following approximations:

(i) the intensity of the incident beam is the same throughout the crystal;

(ii) loss of intensity through rescattering of scattered waves is negligible;

(iii) the distance from P to the crystal is so large in comparison with the dimensions of the crystal that the relative amplitudes at P of the spherical waves from different atoms are not affected significantly by slight differences in their distances from P;

(iv) the scattered waves from different atoms are all nearly parallel at P and approximate to plane waves with the same wave vector, $\mathbf{k}'$.

(i) and (ii) constitute the kinematical approximation, (iii) and (iv) the Fraunhofer approximation. The validity of these approximations will be considered when discussing the dynamical theory.

With these approximations we can write for the contribution to the complex amplitude at P of the atom at $\mathbf{r}_j + \rho_t$,

$$dA = \text{const.} f_t \exp\left[2\pi i(\mathbf{r}_j + \rho_t) \cdot (\mathbf{k}' - \mathbf{k})\right]$$

where f_t is the atomic scattering factor for atoms of type t and the constant depends on the amplitude of the incident wave, the distance from the crystal to P and the absolute scattering power of the atom.

The resultant amplitude at P will therefore be

$$A = \text{const.} \sum_j \sum_t f_t \exp\left[2\pi i(\mathbf{r}_j + \rho_t) \cdot \mathbf{K}\right]$$

where $\mathbf{K} = \mathbf{k}' - \mathbf{k}$ as before and the summations are taken over the whole crystal.

4.3c. *The structure factor*

The general expression for the amplitude at P can be written in the form

$$A = \text{const.} \sum_j \sum_t f_t \exp(2\pi i \mathbf{r}_j \cdot \mathbf{K}) \exp(2\pi i \rho_t \cdot \mathbf{K})$$

$$= \text{const.} \sum_j \left(\sum_t f_t \exp(2\pi i \rho_t \cdot \mathbf{K})\right) \exp(2\pi i \mathbf{r}_j \cdot \mathbf{K})$$

$$= \text{const.} F_K \sum_j \exp(2\pi i \mathbf{r}_j \cdot \mathbf{K}) \tag{4.2}$$

where

$$F_K = \sum_t f_t \exp(2\pi i \rho_t \cdot \mathbf{K})$$

since this is independent of j. F_K depends only on the arrangement of atoms within a unit cell and is known as the structure factor. It describes how the amplitude of the resultant wave from a single unit cell varies with angle and is thus the three-dimensional analogue of the single slit diffraction pattern in one dimension.

We can only observe the modulus of F_K, never its phase, and that only at angles such that a strong diffracted beam is produced by the crystal as a whole. At these angles $\mathbf{K}$ will be a vector of the form $(h\mathbf{b}_1 + k\mathbf{b}_2 + l\mathbf{b}_3)$ with h, k and l integral. If we resolve ρ_t along the axes of the unit cell and write

$$\rho_t = u_t \mathbf{a}_1 + v_t \mathbf{a}_2 + w_t \mathbf{a}_3,$$

then
$$F_{hkl} = \sum_t f_t \exp\left[2\pi i(hu_t + kv_t + lw_t)\right]. \qquad (4.3)$$

An important simplification is possible if the crystal structure possesses a centre of symmetry for then, taking the unit cell such that there is a centre of symmetry at its origin, the atoms within the unit cell can be grouped in pairs with coordinates (u_1, v_1, w_1), $(1-u_1, 1-v_1, 1-w_1)$, etc., and

$$F_{hkl} = \sum_t f_t \cos\{2\pi(hu_t + kv_t + lw_t)\}$$
$$+ i \sum_t f_t \sin\{2\pi(hu_t + kv_t + lw_t)\}$$
$$= \sum_t f_t \cos\{2\pi(hu_t + kv_t + lw_t)\},$$

the sine terms cancelling in pairs. This shows that under these conditions the phase of F_{hkl} can only be 0 or π, and considerably simplifies the calculation of $|F_{hkl}|^2$ and, more important, the phase.

When the lattice planes in a crystal are indexed on the basis of a non-primitive unit cell, as is always done if the symmetry of the lattice is higher than that of any primitive unit cell, then the true nature of some of the lattice points is ignored and the atoms associated with them are treated simply as additional atoms within the larger unit cell. The reciprocal lattice corresponding to this larger unit cell will contain more points than the true reciprocal lattice, but strong diffracted beams will only be observed if $\mathbf{K}$ is a vector of the true reciprocal lattice. The spurious additional points in the reciprocal lattice can be detected by evaluating the structure factor and deleting points for which it is zero. For example, if a structure based on a body-centred cubic lattice and with one atom per lattice point be indexed on a cubic unit cell, then, taking the origin at an atom (which is, of course, a centre of symmetry), the atom positions within the unit cell will be (0, 0, 0) and $(\frac{1}{2}, \frac{1}{2}, \frac{1}{2})$ and the structure factor for the *hkl* reflection will be

$$F_{hkl} = f(1 + \cos\pi(h+k+l))$$
$$= 2f \quad \text{if } (h+k+l) \text{ is even,}$$
$$= 0 \quad \text{if } (h+k+l) \text{ is odd.}$$

If all reciprocal lattice points for which $(h+k+l)$ is odd be deleted, then the remainder will constitute a face-centred cubic lattice, which is the true reciprocal lattice to the body-centred cubic lattice.

This method of detecting spurious points in the reciprocal lattice may only be applied to structures in which there is only one atom per lattice point. With more complex structures it is quite possible for F_{hkl} to be zero at true reciprocal lattice points: deleting these would leave a set of points that did not constitute a lattice at all, just as the atom positions in the structure do not constitute a lattice.

4.3d. *The interference function*

The second term in eq. (4.2) does not depend on the atomic arrangement within a unit cell but instead is determined by the shape and size of the crystal. It can only be evaluated in a simple form for certain simple crystal shapes. For example, for a crystal in the form of a parallelepiped with edges parallel to the axes of a unit cell and of lengths N_1a_1, N_2a_2 and N_3a_3 we can write

$$\mathfrak{I} = \sum_j \exp(2\pi i \mathbf{r}_j . \mathbf{K})$$

$$= \sum_{n_1=1}^{N_1} \sum_{n_2=1}^{N_2} \sum_{n_3=1}^{N_3} \exp[2\pi i(n_1\mathbf{a}_1 + n_2\mathbf{a}_2 + n_3\mathbf{a}_3) . \mathbf{K}]$$

$$= \sum_{n_1=1}^{N_1} \exp(2\pi i n_1\mathbf{a}_1 . \mathbf{K}) \sum_{n_2=1}^{N_2} \exp(2\pi i n_2\mathbf{a}_2 . \mathbf{K}) \sum_{n_3=1}^{N_3} \exp(2\pi i n_3\mathbf{a}_3 . \mathbf{K}).$$

The separation of a triple summation into the product of three summations in this way is only possible for this particular crystal shape, since only for this shape are the limits of each summation independent of the values of the other variables.

The value of $\mathfrak{I}$ will only be appreciable if $\mathbf{K}$ is nearly equal to a reciprocal lattice vector. If we put $\mathbf{K} = \mathbf{g} + \mathbf{s}$, where $\mathbf{g}$ is the vector $(h\mathbf{b}_1 + k\mathbf{b}_2 + l\mathbf{b}_3)$ and $\mathbf{s} = (s_1\mathbf{b}_1 + s_2\mathbf{b}_2 + s_3\mathbf{b}_3)$, then

$$\mathfrak{I}_1 = \sum_{n_1=1}^{N_1} \exp(2\pi i n_1\mathbf{a}_1 . \mathbf{K})$$

$$= \sum_{n_1=1}^{N_1} \exp[2\pi i(n_1h + n_1s_1)]$$

$$= \sum_{n_1=1}^{N_1} \exp(2\pi i n_1 s_1)$$

$$= \frac{\sin(\pi N_1 s_1)}{\sin(\pi s_1)}.$$

C*

The summation is the same as was encountered in the theory of a one-dimensional diffraction grating.

The intensity of the diffracted beam at the point P is therefore

$$J = \text{const.} \, | F_{hkl} |^2 I \qquad (4.4a)$$

where

$$I = |\mathfrak{J}|^2 = \frac{\sin^2 (\pi N_1 s_1)}{\sin^2 (\pi s_1)} \frac{\sin^2 (\pi N_2 s_2)}{\sin^2 (\pi s_2)} \frac{\sin^2 (\pi N_3 s_3)}{\sin^2 (\pi s_3)} . \qquad (4.4b)$$

I is known as the interference function for this crystal shape. Note that it is independent of $\mathbf{g}$.

The constant in eq. (4.4a), which we shall call B, has the value (I_0/R^2) for electrons and neutrons, where I_0 is the intensity of the incident beam and R is the distance from the crystal to the point P. For X-rays the value of B is $(I_0/R^2)(e^2/mc^2)^2 \sin^2 \chi$, where χ is the angle between the electric field in the incident beam and the plane containing the incident and diffracted beams, the difference arising through the different definition of f in this case. For an unpolarised beam of X-rays,

$$B = (I_0/R^2)(e^2/mc^2)^2 \tfrac{1}{2}(1 + \cos^2 2\theta).$$

J can be thought of as a scattering function in reciprocal space, the value of this function at any point, M, on the reflecting sphere giving the intensity of the diffracted beam in the direction normal to the sphere at that point. J will only be appreciable within a small volume of reciprocal space around a reciprocal lattice point. Within this volume the variation of J will be due almost entirely to the variation of I, and so will be the same around every reciprocal lattice point.

Calculation of the exact form of I for a particular crystal shape is not an easy matter. The general features, however, can be inferred from a consideration of diffraction by planar apertures: the interference function will be extended in reciprocal space parallel to directions in which the crystal is narrow, and vice versa, as can be seen for the case of a parallelepiped. Some examples of crystal shapes and the corresponding interference functions are shown in Fig. 4.4.

4.3e. *The integrated intensity*

J gives the intensity of the diffracted wave at P. The total rate at which energy is being received at a counter or a spot on a film will be

$$P = \int J \, dS$$

CRYSTAL SHAPE   INTERFERENCE FUNCTION

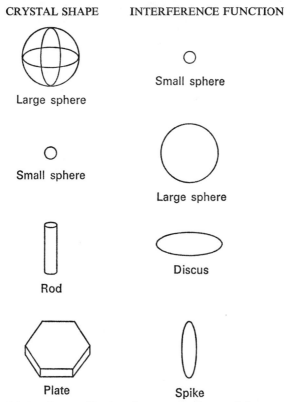

Large sphere

Small sphere

Small sphere

Large sphere

Rod

Discus

Plate

Spike

FIG. 4.4. Interference functions for some simple crystal shapes.

where dS is an element of area at P normal to the diffracted beam, and the integral is taken over the aperture of the counter or the area of the spot. The total energy received at the detector in time t will be

$$E = \int_0^t P \, dt.$$

This can be evaluated most simply by transforming it into an integral in reciprocal space. For the X-ray case we have

$$E = \int_0^t \left(\int J \, dS \right) dt$$

$$= \int_0^t \left(\int \frac{I_0}{R^2} \left(\frac{e^2}{mc^2} \right)^2 \tfrac{1}{2}(1 + \cos^2 2\theta) |\, F_K\,|^2 I \, dS \right) dt.$$

Since I is only appreciable over a very small range of K, we may replace F_K by its value, F_{hkl}, at the centre of this range and take it outside the integral: likewise for $\cos^2 2\theta$. Then

$$E = I_0 \left(\frac{e^2}{mc^2}\right)^2 \tfrac{1}{2}(1+\cos^2 2\theta)|\, F_{hkl}\, |^2 \int_0^t (\int I\, d\Omega)\, dt$$

where $d\Omega = dS/R^2$ is the solid angle subtended by the area dS at the crystal.

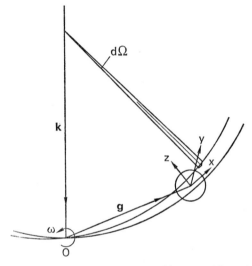

Fig. 4.5. Geometry of calculation of integrated intensity.

For a crystal that is being rotated with uniform angular velocity ω about an axis normal to the plane containing $\mathbf{k}$ and $\mathbf{g}$, the reflecting sphere will sweep through the reciprocal lattice point defined by $\mathbf{g}$ moving very nearly parallel to itself with velocity $g\omega \cos \theta$. If we take rectangular axes in reciprocal space with the origin at the reciprocal lattice point, the z-axis normal to the reflecting sphere and the x- and y-axes in the tangent plane to the reflecting sphere at the reciprocal lattice point (Fig. 4.5), then

$$I(s_1, s_2, s_3) = I'(x, y, z),$$

$$d\Omega = \frac{dx\, dy}{k^2},$$

$$dz = g\omega \cos \theta \, dt = k\omega \sin 2\theta \, dt$$

and

$$\int_0^t (\int I \, d\Omega) \, dt = \iiint I' \frac{dx \, dy}{k^2} \frac{dz}{k\omega \sin 2\theta}$$

$$= \frac{\lambda^3}{\omega \sin 2\theta} \iiint I' \, dx \, dy \, dz,$$

$dx \, dy \, dz$ is an element of volume in reciprocal space. The value of the integral will be the same for every reciprocal lattice point and will give what might be considered as the total mass of the distribution, the interference function being analogous to a density function.

Evaluation of the integral will be a straightforward matter if I' is known as a function of x, y and z. For the case of a parallelepipedal crystal it will be convenient to take as our element of volume in reciprocal space the parallelepiped with edges parallel to the reciprocal lattice axes and of lengths $ds_1 \mathbf{b}_1$, $ds_2 \mathbf{b}_2$ and $ds_3 \mathbf{b}_3$. The volume of this element will be

$$dv^* = (ds_1 \mathbf{b}_1 \wedge ds_2 \mathbf{b}_2) \cdot ds_3 \mathbf{b}_3$$

$$= ds_1 ds_2 ds_3 (\mathbf{b}_1 \wedge \mathbf{b}_2) \cdot \mathbf{b}_3$$

$$= ds_1 ds_2 ds_3 v^*$$

$$= ds_1 ds_2 ds_3 / v$$

where v^* and v are the volumes of the unit cells of the reciprocal and crystal lattices respectively. Then

$$\iiint I' \, dx \, dy \, dz = \frac{1}{v} \iiint I \, ds_1 ds_2 ds_3.$$

I is given by eq. (4.4b) and the triple integral can be expressed as the product of three integrals each of the form

$$\int \frac{\sin^2 (\pi N_1 s_1)}{\sin^2 (\pi s_1)} \, ds_1.$$

The limits of the integral will depend on the angular range of rotation of the crystal and the aperture of the detector, but provided these are sufficient to cover the range of values in which the integrand is significantly different from zero the exact values are immaterial. Within this range s_1 is very small and so

$$\int \frac{\sin^2 (\pi N_1 s_1)}{\sin^2 (\pi s_1)} \, ds_1 \simeq \frac{1}{\pi} \int_{-\infty}^{\infty} \frac{\sin^2 (\pi N_1 s_1)}{(\pi s_1)^2} \, d(\pi s_1)$$

$$= N_1,$$

$$\therefore \quad \frac{1}{v} \iiint I \, ds_1 ds_2 ds_3 = N_1 N_2 N_3 / v$$

$$= V/v^2 = n^2 V$$

where V is the volume of the crystal and $n = 1/v$ is the number of unit cells per unit volume. Finally,

$$\frac{E\omega}{I_0} = \left(\frac{e^2}{mc^2}\right)^2 \tfrac{1}{2}(1+\cos^2 2\theta)| F_{hkl} |^2 \frac{\lambda^3}{\sin 2\theta} n^2 V$$

$$= QV$$

where

$$Q = \frac{n^2 \lambda^3}{\sin 2\theta} \left(\frac{e^2}{mc^2}\right)^2 \tfrac{1}{2}(1+\cos^2 2\theta)| F_{hkl} |^2.$$

The quantity QV is known as the integrated intensity of reflection. In the general case of a crystal of arbitrary shape it is again found to be proportional to the volume of the crystal.

4.3f. *Thermal motions*

So far we have considered only scattering centres at, or in a fixed position relative to, the points of a lattice. In practice this ideal is never realised since the atoms in any structure vibrate about their mean positions with an amplitude that depends on the temperature. This leads to a reduction in the amplitude of the diffracted beam at the Bragg angle and an increase in the background scattering away from the Bragg angle. The effect on the integrated intensity can be described by a *temperature factor*. The magnitude of the effect can be calculated quite simply for a structure with only one atom per lattice point, as follows.

At the Bragg angle in the absence of thermal vibrations the amplitude–phase diagram would be a straight line, all the scattered waves being exactly in phase. If the atom positions are affected by thermal vibrations, then there will be small variations of phase and the amplitude–phase diagram will be as shown in Fig. 4.6. At a later time all the phases will have altered but the general appearance of the amplitude–phase diagram will be the same. (This method of treatment is possible because the frequency of vibration of an atom is very much less than the frequency of

FIG. 4.6. Amplitude–phase diagram at the Bragg angle when the atoms are displaced slightly from their mean positions, on account of thermal vibrations.

the incident radiation, and so over a comparatively long period of time the phases will be substantially constant.) The resultant amplitude is

$$R = \sum_j f_j \cos \varepsilon_j,$$

the summation being taken over all atoms. Provided that the phase angles, ε_j, are small this can be written

$$R = \sum_j f_j (1 - \tfrac{1}{2}\varepsilon_j^2 + \ldots)$$
$$\simeq Nf(1 - \tfrac{1}{2}\overline{\varepsilon^2})$$
$$\simeq Nf e^{-\frac{1}{2}\overline{\varepsilon^2}}$$
$$= Nf e^{-M}$$

where N is the total number of atoms, each of scattering power f, $\overline{\varepsilon^2}$ denotes the mean value of ε_j^2 and $M = \tfrac{1}{2}\overline{\varepsilon^2}$. The amplitude of the resultant is therefore reduced by the factor e^{-M}. M can be related to the mean square displacement of an atom from its mean position, which we take as the lattice point. If u_j is the displacement of atom j normal to the operative set of reflecting planes (of spacing d), then

$$\varepsilon_j = 2\pi \frac{u_j}{d}$$
$$= 4\pi u_j \sin \theta / \lambda$$

and

$$M = 8\pi^2 \left(\frac{\sin^2 \theta}{\lambda^2} \right) \overline{u^2}.$$

Since the amplitude is reduced by a factor e^{-M} the intensity at any angle close to the Bragg angle, and therefore also the integrated intensity, is reduced by a factor e^{-2M}. The situation is formally equivalent to replacing f by fe^{-M}. Note that M depends on the structure and the temperature, is different for different atoms in the same structure, and may even be different for the same atoms in different reflections.

Away from the Bragg angle cancellation of waves from different atoms is no longer complete as a result of thermal vibrations, and so the background intensity is increased. This would not happen with a set of fixed atoms of scattering power fe^{-M} situated at the lattice points. For an adequate treatment of the background scattering we would have to take into account the fact that the atom displacements are not independent of one another. This will not be attempted.

4.4. Dynamical theory of diffraction

4.4a. *Introduction*

A rigorous treatment of diffraction by space lattices would involve establishing and then solving, with the appropriate boundary conditions, the equation of wave motion in the field of the lattice, and is beyond the scope of this book. Instead, we shall follow a wave optical treatment that is closely similar to the treatment of earlier chapters. This cannot be considered rigorous, for reasons that will be indicated: its justification lies in the fact that it leads to results that are identical with those obtained by a more rigorous treatment.

Two extreme cases can be distinguished, in which the reflecting planes are parallel and perpendicular respectively to the crystal surface. The formulation of the equations for the amplitudes of the transmitted and diffracted beams are very similar in the two cases, but the developments are rather different, reflecting different interests. In the parallel, or Bragg, case the chief interest is in the intensity of the diffracted beam from a large, perfect crystal, whereas in the perpendicular, or Laue, case the interest is in diffraction by small crystals containing imperfections and in the identification of those imperfections.

4.4b. *The refractive index of a crystal*

We saw in Chapter 3 that when a beam of radiation is incident on a plane of atoms there is a forward scattered beam that is shifted in phase by $\pi/2$ relative to the incident beam. The resultant wave incident on the next plane of atoms is therefore shifted slightly in phase and so the wavelength in the crystal, which is the distance between planes in which the disturbance has the same phase, is altered. We can calculate the value of the refractive index of the crystal as follows.

Suppose that successive atom planes in the crystal, parallel to the surface, are separated by a distance δz. For a wave of amplitude A incident normally on the surface, the amplitude of the forward scattered wave from each plane is $iq_0 A$, where

$$q_0 = n\lambda a f(0).$$

For electrons $a = 1$; for X-rays $a = -e^2/(mc^2)$, the minus sign indicating that the scattered wave from each atom is exactly out of phase with the incident wave.

The amplitude and phase of the wave at the next atom plane will therefore be

$$A(1+iq_0) \exp(2\pi ik\,\delta z) = A \exp(iq_0) . \exp(2\pi ik\,\delta z)$$

$$= A \exp[2\pi i(k + N\lambda a f(0)/2\pi)]\,\delta z,$$

where $N = n/\delta z$ is the number of atoms per unit volume. For complex structures $f(0)$ should be replaced by F_{000}, the zero order structure factor for a unit cell, and N becomes the number of unit cells per unit volume.

The effective value of the wave vector is now

$$k^* = k + \frac{N\lambda a f(0)}{2\pi}$$

and so the refractive index is

$$\mu = \frac{k^*}{k} = 1 + \frac{N\lambda^2 a f(0)}{2\pi} .$$

For X-rays this becomes

$$\mu = 1 - \frac{N\lambda^2 e^2}{2\pi mc^2} f(0).$$

This is always less than but very close to unity. In consequence, X-rays can be totally reflected from the surface of a crystal at glancing angles of incidence. For copper metal and X-rays of wavelength $\lambda = 1\cdot54$ Å (Cu Kα radiation), $\mu = 1 - 0\cdot000027$, and so the effects of refraction on entering a specimen can generally be neglected.

In deriving eq. (3.2) we assumed that there were sufficient atoms in the first Fresnel zone for them to be considered as continuously distributed. This assumption was valid when the amplitude of the scattered wave at some distance from the atom plane was being considered: here, however, we have assumed that the same equation for the scattered amplitude holds very close to the atom plane, in fact at a distance where the number of atoms in a Fresnel zone may not even be as great as one. The justification of this assumption lies in the results derived, as already mentioned.

4.4c. *The Bragg case*

Let the atom planes parallel to the surface of the crystal be denoted by a serial number 0, 1, 2, ..., j, ... and let the amplitudes of the transmitted and diffracted waves immediately above the jth plane be denoted by T_j and S_j respectively (Fig. 4.7a).

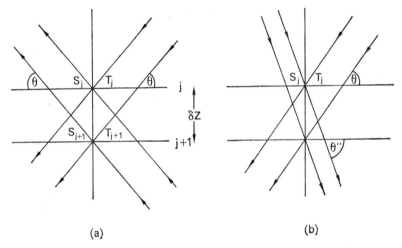

(a)   (b)

Fig. 4.7. Interaction of transmitted and reflected waves at each atom plane:
(a) Bragg case; (b) Laue case.

Since the dimensions of each plane are very large in comparison with the wavelength of the radiation, the amplitude of the diffracted wave will only be appreciable in a direction closely satisfying the law of reflection for each plane. However, the angles of incidence and reflection will not necessarily satisfy the Bragg law for reflection from this set of planes. The phase change, α, in either wave in traversing the distance $\delta z \sin \theta$ between successive planes will be

$$\alpha = 2\pi k \, \delta z \sin \theta.$$

Now T_{j+1} will consist of T_j together with the forward scattered wave from plane j, both of which will have traversed a distance $\delta z \sin \theta$ and suffered a phase change of α relative to T_j, and also the part of S_{j+1} which has been reflected from the underside of plane j and which will have suffered a phase change of 2α relative to S_{j+1}. Therefore, if we assume the same reflection coefficient, iq, for reflection from the two sides of a plane (Friedel's law)

$$T_{j+1} = (1+iq_0)T_j e^{i\alpha} + iqS_{j+1}e^{2i\alpha}. \tag{4.5}$$

Also, by similar arguments,

$$S_j = iqT_j + (1+iq_0)S_{j+1}e^{i\alpha}. \tag{4.6}$$

We will try as a solution of this set of equations

$$T_{j+1} = \chi T_j. \tag{4.7}$$

Then, from (4.5),

$$\chi(T_j - (1+iq_0)T_{j-1}e^{i\alpha}) = iqS_{j+1}e^{2i\alpha}.$$

But, reducing the index of (4.5) by unity,

$$T_j - (1+iq_0)T_{j-1}e^{i\alpha} = iqS_j e^{2i\alpha}. \tag{4.5*}$$

Therefore

$$S_{j+1} = \chi S_j$$

and so, from (4.6),

$$\frac{S_j}{T_j} = \frac{iq}{1 - \chi(1+iq_0)e^{i\alpha}}. \tag{4.8}$$

χ can be obtained by eliminating S between eqs. (4.5) and (4.6) and reducing the index of T to the same value throughout by use of eq. (4.7). If we substitute for S_j in eq. (4.5*) the value given by eq. (4.6) and for S_{j+1} in eq. (4.6) the value given by eq. (4.5) we have

$$\frac{T_j - (1+iq_0)T_{j-1}e^{i\alpha}}{iqe^{2i\alpha}} = iqT_j + (1+iq_0)e^{i\alpha}\left\{\frac{T_{j+1} - (1+iq_0)T_j e^{i\alpha}}{iqe^{2i\alpha}}\right\}.$$

Rearranging,

$$(1+iq_0)(T_{+1} + T_{j-1})e^{i\alpha} = (q^2 e^{2i\alpha} + (1+iq_0)^2 e^{2i\alpha} + 1)T_j.$$

Dividing by $T_j e^{i\alpha}$ and using eq. (4.7),

$$(1+iq_0)(\chi+(1/\chi)) = q^2 e^{i\alpha}+(1+iq_0)^2 e^{i\alpha}+e^{-i\alpha}. \tag{4.9}$$

This equation for χ can be simplified. We expect that T will not vary much from one plane to the next apart from the phase factor, which is nearly equal to π for angles of incidence that nearly satisfy the Bragg law, and so we put

$$\chi = (1-\chi^*)e^{i\pi} = -(1-\chi^*).$$

χ^* may be complex, which takes account of any phase difference of χ from π. α also is very nearly equal to π for angles of incidence close to the Bragg angle, so we put

$$\alpha = \pi+\varepsilon, \qquad e^{i\alpha} = -e^{i\varepsilon}.$$

Then, expanding eq. (4.9) and neglecting cubes and higher powers of small quantities, we find

$$\chi^{*2} = q^2-(q_0+\varepsilon)^2.$$

The ratio of the diffracted and incident amplitudes is given by eq. (4.8) with $j = 0$. Substituting for χ and α, expanding $e^{i\varepsilon}$ and retaining only the first powers of small quantities in the expansion of the denominator we find

$$\frac{S_0}{T_0} = \frac{iq}{\chi^*-i(q_0+\varepsilon)}$$

$$= \frac{-q}{(q_0+\varepsilon)\pm i(q^2-(q_0+\varepsilon)^2)^{\frac{1}{2}}}.$$

$|S_0/T_0|^2$ is plotted as a function of ε, which is a measure of the deviation from the exact Bragg angle, in Fig. 4.8, which is drawn for the case of no absorption (q and q_0 real). It will be seen that total reflection occurs over a narrow range of angles centred on the angle for which $\varepsilon = -q_0$, and that the angular width of the range of total reflection is equal to $2q$. Outside this range of angles the intensity of the diffracted beam falls rapidly.

Since $q < q_0$, $\varepsilon = 0$ does not even fall inside the range of total reflection. At first this may seem surprising until it is realised that all the angles involved are very small and that the refractive index effect, small as it is, changes the wavelength of the radiation inside the crystal by just a sufficient amount to satisfy the Bragg law exactly at the orientation of the crystal for which $\varepsilon = -q_0$, i.e. at the centre of the range of total reflection.

The integrated intensity is obtained by integrating the intensity of reflection over the angular width of the reflection, as before. When this is done it is found that the wings of the intensity distribution contribute one-third as much to the integrated intensity as does the region of total reflection and that therefore the integrated intensity is proportional to the angular width of this region, i.e. to q and hence to $|F_{hkl}|$. Note also that, under the conditions of this section, the integrated intensity does not depend on the thickness of the crystal, nor, therefore, on its volume. The implications of these results will be discussed in Section 4.5.

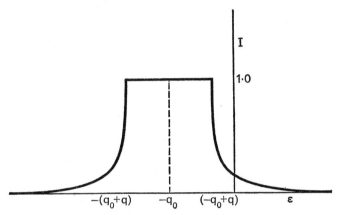

FIG. 4.8. Reflection curve for a large, perfect crystal with no absorption: drawn for a reflection such that $q = \frac{2}{3}q_0$.

4.4d. *The Laue case*

When considering the Bragg case we supposed that only the "reflected" waves from the set of planes parallel to the surface of the crystal interfered constructively, i.e. that the crystal was in an orientation nearly satisfying the Bragg condition for this set of planes but for no other. However, as we saw in Section 3.4c, there will be many other diffracted waves formed from a single plane of atoms in addition to the forward scattered and reflected waves. We will now consider the case in which one of these waves, but not the reflected wave, is reinforced by constructive interference within the crystal. Whenever constructive interference occurs the diffracted beam can be thought of as having been reflected from one of the many sets of planes in the crystal. In the Laue case we suppose that this set of planes is normal to the surface of the crystal. We will retain the notation of Fig. 3.7, but it should be noted that θ is now not the Bragg angle for the reflection being considered. In fact, for the Laue case, $\theta \simeq \theta'' \simeq \pi/2 - \theta_B$, where θ_B, the angle between the incident beam and the reflecting planes at the reflecting orientation, is the Bragg angle.

As before, we will denote the atom planes parallel to the surface by a serial number $0, 1, 2, \ldots, j, \ldots$, and let the amplitudes of the transmitted and diffracted waves immediately above the jth plane be denoted by T_j and S_j respectively (Fig. 4.7b). Since the dimensions of each plane are very large in comparison with the wavelength of the radiation, the amplitude of the diffracted wave will only be appreciable in a direction closely satisfying the equation $d(\cos \theta + \cos \theta'') = p\lambda$. The angles θ and θ'' will in general not be equal and so the phase changes in the transmitted and diffracted waves in traversing the distance between successive planes will not be the same. However, since the intensity of the diffracted wave will only become appreciable if the Bragg condition is nearly satisfied, i.e. if $\theta \simeq \theta''$, we can put

$$\theta = \pi/2 - \theta_B + \delta\theta,$$
$$\theta'' = \pi/2 - \theta_B - \delta\theta,$$

where $\delta\theta$ is the angular deviation from the exact reflecting position. The phase changes between planes in the transmitted and diffracted waves will then be

$$\alpha_T = 2\pi k \delta z \cos (\theta_B - \delta\theta)$$

and

$$\alpha_S = 2\pi k \delta z \cos (\theta_B + \delta\theta)$$

respectively, and the equations of propagation become

$$T_{j+1} = \{(1 + iq_0)T_j + iqS_j\}e^{i\alpha_T},$$
$$S_{j+1} = \{(1 + iq_0)S_j + iqT_j\}e^{i\alpha_S}.$$

T_{j+1} will differ from T_j chiefly through the propagation phase factor $e^{i\alpha_T}$. Since the relative phases at different atomic planes are of no great interest to us (unlike the situation in the Bragg case), it will be convenient to define new quantities T' and S' such that

$$T'_j = T_j e^{-ij\alpha_T},$$
$$S'_j = S_j e^{-ij\alpha_T}.$$

This merely represents a change in the origin of phase from one atomic plane to the next: the relative phases of T' and S' at any point in the crystal are the same as those of T and S. Then

$$T'_{j+1} = (1+iq_0)T'_j+iqS'_j, \tag{4.10a}$$

$$S'_{j+1} = \{(1+iq_0)S'_j+iqT'_j\}e^{i(\alpha_S-\alpha_T)}. \tag{4.10b}$$

The phase factor $e^{i(\alpha_S-\alpha_T)}$ represents the relative change in phase of T and S between planes due to the different distances travelled, and is small. It is related to the distance of the reflecting sphere from the appropriate reciprocal lattice point, s_g, as follows:

$$\alpha_S-\alpha_T = 2\pi k \, \delta z \, . \, 2 \sin \theta_B \, \delta\theta$$

$$= 4\pi k \sin \theta_B \, \delta\theta \, . \, \delta z.$$

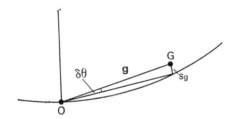

Fig. 4.9. The relation between the deviation parameter, s_g, and the angular deviation, $\delta\theta$, from the Bragg angle.

Now, in Fig. 4.9,

$$s_g = g \, \delta\theta$$

$$= 2k \sin \theta_B \, \delta\theta.$$

Therefore

$$\alpha_S-\alpha_T = 2\pi s_g \, \delta z.$$

We adopt the convention that s_g is positive if the reciprocal lattice point lies inside the reflecting sphere, as in Fig. 4.9.

Returning to eq. (4.10), putting $T'_{j+1} = T'_j+\delta T'$, $S'_{j+1} = S'_j+\delta S'$, dropping the suffix and expanding the phase factor, we have

$$\delta T' = iq_0 T'+iqS',$$

$$\delta S' = iqT'+i(q_0+2\pi s \, \delta z)S'$$

whence

$$\frac{dT'}{dz} = \frac{i\pi T'}{\xi_0} + \frac{i\pi S'}{\xi_g}, \tag{4.11a}$$

$$\frac{dS'}{dz} = \frac{i\pi T'}{\xi_g} + i\left\{\frac{\pi}{\xi_0} + 2\pi s\right\}S' \tag{4.11b}$$

where

$$\frac{q_0}{\delta z} = \frac{\pi}{\xi_0}; \qquad \frac{q}{\delta z} = \frac{\pi}{\xi_g}.$$

ξ_0 and ξ_g are to be regarded as constants whose significance will become apparent later.

This pair of coupled differential equations can be simplified and solved explicitly and it is instructive, but beyond the scope of this book, to do so. They can also, of course, be solved by numerical methods, with the aid of a computer, and it is for this reason that the original equations have been reduced to this form. The results obtained are shown in Fig. 4.10 for various values of s.

It will be seen that ξ_g is equal to the distance in which T' (or S') undergoes a complete cycle of intensity variation when the crystal is oriented at the exact reflecting position. It is known as the extinction distance. For diffraction of electrons ξ_g is given by

$$\xi_g = \frac{\pi \, \delta z \sin \theta}{n\lambda f(\theta_B)} .$$

For complex structures this becomes

$$\xi_g = \frac{\pi \sin \theta}{N\lambda F_{hkl}} = \frac{\pi V \cos \theta_B}{\lambda F_{hkl}}$$

where $N = n/\delta z$ is the number of unit cells per unit volume, $V = 1/N$ is the volume of a unit cell and F_{hkl} is the structure factor for the reflecting planes, of indices hkl.

The significance of ξ_0 can be seen by considering the transmitted wave in a region where $S' = 0$. Equation (4.11) then becomes

$$\frac{dT'}{dz} = \frac{i\pi T'}{\xi_0}.$$

This shows that the phase of T varies slowly with depth in the crystal from that of a wave with phase propagation factor $e^{i\alpha_T z}$; in other words, the wave propagates with a wave vector that is different from k by an amount depending on ξ_0, which is therefore related to the refractive index of the crystal.

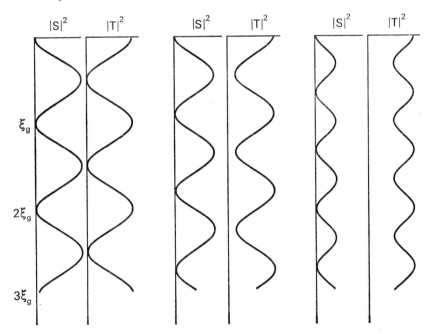

FIG. 4.10. The variation of intensity of transmitted and diffracted beams with depth in a crystal. The curves are drawn for the 111 reflection in copper of 100 keV electrons: (a) $\delta\theta = 0$; (b) $\delta\theta = 4 \times 10^{-3}$; (c) $\delta\theta = 10^{-2}$. Similar curves would occur at smaller deviations with weaker reflections.

4.5. Real crystals

We will now consider to what extent the theories developed in this chapter apply to real crystals. The most obvious difference between the ideal crystals considered so far and real crystals is that all real crystals show absorption—that is, energy is lost from a beam of radiation by incoherent, inelastic scattering processes (see Chapter 5), and also, particularly with X-rays, by atomic absorption. Consequently the radiation incident on an atom at some depth in a crystal or polycrystalline specimen is less intense than the radiation at the surface. This can be allowed for quite simply if the variation in intensity of incident radiation through the volume of a crystallite is small, for then the kinematical theory as developed in Section 4.3 is a good approximation to reality, and in particular the integrated intensity from a crystallite is proportional to its volume. The exact form of the correction to be applied depends on the geometry of the arrangement.

If we inquire how large a crystallite may be for the kinematical theory still to be a good approximation, we find that the reduction of intensity with depth due to absorption is generally negligible in comparison with the reduction due to dynamical effects, as a short calculation will show. For the Bragg case of the dynamical theory we have $T_{j+1} = \chi T_j$, and at the modified Bragg angle (modified to allow for the refractive index effect), $|\chi| = |1-\chi^*| = 1-q$. The transmitted beam is therefore attenuated exponentially with depth in the crystal. This attenuation is known as primary extinction, to distinguish it from the related phenomenon of secondary extinction, which will be considered later, and from true absorption. Now

$$|T_j| = |\chi|^j T_0 = \exp(-\mu^* j) T_0,$$

$$|T_j|^2 = \exp(-2\mu^* l \sin\theta/\delta z) |T_0|^2$$

$$= \exp(-\mu' l) |T_0|^2$$

where $\mu^* = -\ln|\chi| = q$, l is the actual distance travelled by the beam in the crystal and $\mu' = 2\mu^* \sin\theta/\delta z$ is an effective linear absorption coefficient due to primary extinction. For copper metal and X-rays of wavelength $1{\cdot}54$ Å, $\mu' = 16{,}000$ cm^{-1} for the low-order reflections (depending on $f(\theta)$), though, of course, the attenuation is much less than this away from the exact reflecting position, and is also less for weaker reflections. For comparison, the ordinary linear absorption coefficient for the same conditions is 470 cm^{-1}. Attenuation curves for various low-order reflections in copper are shown in Fig. 4.11, together with true absorption curves. As can be seen, primary extinction is significant for the low-order reflections for crystallites of size $0{\cdot}5$ μm, and yet it is found experimentally that the kinematical theory is often satisfactory for much larger crystallites: the intensities of the lines on a powder diffraction photograph from annealed copper filings agree well with the predictions of the theory.

The explanation of this is that what appears at first sight to be a large, perfect crystal is often very far from the degree of perfection required for the dynamical theory to apply. Originally it was supposed that a crystal would consist of small blocks slightly misoriented with respect to one another, the arrangement being known as a mosaic. Basically this is still

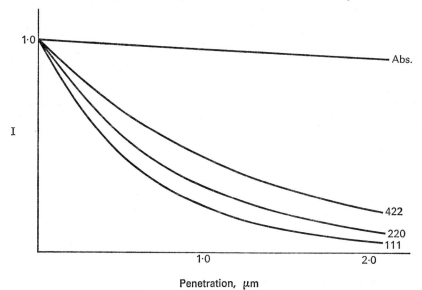

Fɪɢ. 4.11. Extinction and absorption curves in copper for X-rays of wavelength 1·54 Å.

the view, though the boundaries of the blocks are probably not always as clearly defined as was first thought: lattice strains or dislocations would disrupt the exact regularity of the lattice and so cause neighbouring regions of a crystal to reflect essentially independently. Nevertheless, large perfect crystals for which the dynamical theory is applicable do occur, and in electron diffraction, due to the much larger scattering power of an atom for electrons than for X-rays or neutrons, the dynamical theory is nearly always required.

Even if the blocks of a mosaic structure are sufficiently small for primary extinction within a block to be negligible, the fact that radiation is diffracted away by a block will lead to a reduction in the intensity of the radiation incident on any block deeper in the crystal. This is known as secondary extinction. Of course, what matters is not the overall reduction of intensity in the incident beam but the reduction of intensity within the very narrow angular range that the deeper block will reflect. The importance of secondary extinction will therefore depend on the mosaic spread within the crystal, amongst other factors, since this will determine the probability that two blocks will have closely the same orientation, and so may be quite different for what appear to be identical crystals. Secondary extinction can be allowed for by using a modified absorption coefficient when calculating the attenuation of incident or diffracted beams in passing through a crystal, provided that the mosaic blocks are small enough, and therefore sufficiently numerous for statistical fluctuations in their angular distribution to be negligible. Secondary extinction differs from true absorption, however, in that it depends on the intensity of the reflection, and so it is different for different reflections from the same crystal, as well as being different, in general, for the same reflection from different crystals.

In view of what has been said it may seem surprising that reliable and interpretable measurements of integrated intensity can be made at all. Indeed, one should always be on guard against the effects of primary and secondary extinction when comparing intensities of diffracted beams, and work, if possible, with weak reflections, small crystals and radiations of short wavelength.

Absorption also modifies the shape of the intensity versus angle curve in the dynamical theory (Fig. 4.8). As might be expected, absorption leads to a reduction in the intensity of the reflected beam, and this reduction is greater the greater would have been the penetration of the incident beam into the crystal in the absence of absorption. Absorption can be allowed for formally by letting q_0 and q take complex values. It is almost impossible to check in any detail the resulting expressions for the intensity of reflection, and furthermore, a more rigorous treatment of the dynamical theory indicates that the wave optical treatment developed in this chapter is inadequate for a discussion of absorption, so we shall not pursue the subject further.

4.6. Diffraction by imperfect lattices

We will conclude this chapter by considering what modifications are necessary to eq. (4.11) when the lattice contains imperfections such as dislocations. The theory was developed in order to interpret the contrast in electron micrographs from metal foils containing lattice defects and this is still the chief interest. The method of treatment involves approximations that are only valid in this case, and so this section is really only applicable to electron diffraction.

In the electron microscope a beam stop is placed in the plane *FF'* of Fig. 2.18. This allows only one of the beams to pass and form the final image. Contrast arises because the intensity of this beam (usually the zero-order beam) as it emerges from the specimen is not everywhere the same, but depends on the perfection of the lattice in the vicinity of the point of emergence. This in turn leads to a broadening of and a fine structure within the corresponding diffraction spot, on a scale that is inversely proportional to the scale of intensity variations in the object plane.

To treat the problem we make what is known as the column approximation: we suppose that the intensity of any beam at the lower surface of the foil is determined by the incident beam intensity and by diffraction occurring solely within a column in the foil centred on the point of emergence at which the intensity is to be calculated; we suppose further that this column is sufficiently wide for the Fresnel zone treatment of diffraction by a plane of atoms to be valid for each plane in the column, and yet sufficiently narrow for each plane to be displaced, by lattice strains and the like, essentially as a rigid unit. That it is possible to find a column width that satisfies these two conditions is largely due to the very short wavelengths of energetic electrons, as used in an electron microscope. Also, the angle between the transmitted and diffracted beams is very small and so it really makes no difference whether we choose a column that is parallel to either of these or to some other close direction, since the boundaries of the column do not enter into the discussion. We will consider a column that is normal to the plane of the foil, as in Fig. 4.12a.

Consider a plane of atoms that is displaced by a vector **R** from the position that it would occupy in a perfect lattice (Fig. 4.12b). The scattered wave from each atom in this plane will have a phase different from what it would have had had it not been displaced, the phase difference being given by

$$\alpha = 2\pi \mathbf{R} \cdot (\mathbf{k}' - \mathbf{k}).$$

We are only interested in the wave that is scattered in a direction that is close to satisfying the Bragg law, so we can put $\mathbf{k}' - \mathbf{k} \simeq \mathbf{g}$, where $\mathbf{g}$ is a vector of the reciprocal lattice. Then $\alpha = 2\pi\mathbf{g}\cdot\mathbf{R}$. Since all the atoms in a plane parallel to the surface of the crystal are displaced by the same amount, their relative phases will not be altered and so the geometry of the Fresnel zones will be the same as before. The resultant amplitude will therefore also be the same, but its phase will be changed by α. The displacement can therefore be allowed for by replacing q by $qe^{i\alpha}$ in the scattering of the T wave and by $qe^{-i\alpha}$ in the scattering of the S wave, the change of sign being due to the fact that in the latter case the wave is scattered from wave vector $\mathbf{k}'$ to wave vector $\mathbf{k}$. q_0 is left unchanged, since for scattering in

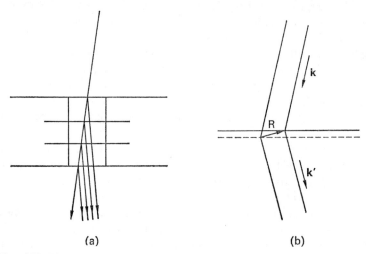

(a)   (b)

Fig. 4.12. The column approximation in transmission electron microscopy: (a) geometry of the column; (b) phase change resulting from displacement of an atom plane by a vector **R**.

the forward direction the initial and final wave vectors are the same, and a small displacement leads to no change in the phase of the scattered wave. With these changes eqs. (4.11) become

$$\frac{dT'}{dz} = \frac{i\pi T'}{\xi_0} + \frac{i\pi S'}{\xi_g}\, e^{i\alpha}, \tag{4.12a}$$

$$\frac{dS'}{dz} = \frac{i\pi T'}{\xi_g}\, e^{-i\alpha} + i\left\{\frac{\pi}{\xi_0} + 2\pi s\right\}S'. \tag{4.12b}$$

These equations are not adequate to describe many of the features of the passage of electrons through crystals, since they do not take absorption into account. As we have seen, absorption can be allowed for formally by letting q_0 and q take complex values, which in this case is equivalent to letting ξ_0 and ξ_g be complex. If we replace $1/\xi_0$ by

$$\left(\frac{1}{\xi_0} + \frac{i}{\xi_0'}\right) \quad \text{and} \quad \frac{1}{\xi_g} \quad \text{by} \quad \left(\frac{1}{\xi_g} + \frac{i}{\xi_g'}\right)$$

the equations predict some anomalous absorption effects which in fact are observed in practice. These effects cannot be interpreted physically in terms of the wave-optical model of electron propagation, but if we return to first principles and attempt to solve the wave equation in the field of the crystal lattice they do receive a very clear and simple explanation. For details see Hirsch *et al.* (1965).

Equations (4.12), modified to allow for absorption, are not in the simplest possible form for numerical computation. To simplify them we make the substitutions

$$T'' = T' \exp(-i\pi z/\xi_0),$$
$$S'' = S' \exp(-i\pi z/\xi_0) \exp(i\alpha).$$

The equations then become

$$\frac{dT''}{dz} = -\frac{\pi T''}{\xi_0'} + \pi\left(\frac{i}{\xi_g} - \frac{1}{\xi_g'}\right) S'',$$

$$\frac{dS''}{dz} = \pi\left(\frac{i}{\xi_g} - \frac{1}{\xi_g'}\right) T'' + \left(-\frac{\pi}{\xi_0'} + 2\pi i(s+\beta)\right) S''$$

where

$$\beta = \frac{1}{2\pi}\frac{d\alpha}{dz} = \mathbf{g}\cdot\frac{d\mathbf{R}}{dz}.$$

Note that T'', T' and T differ only through phase factors, and so $|T''| = |T'| = |T|$. The intensity of the transmitted beam will be equal to the square of the modulus of any one of them.

Usually one loses sight of the physics of a problem when it disappears inside a computer. In this case, however, the final form of the equations helps us to understand some of the features of the contrast. In particular, we should note that it is the local rotation of the lattice towards or away from the exact reflecting position, represented by β in conjunction with the deviation parameter, s, that determines the intensity of diffraction. The actual atomic displacements are of no significance.

CHAPTER 5

SCATTERING PROCESSES

5.1. Introduction

In earlier chapters we have emphasised the similarities between various types of diffraction phenomena. Different techniques are not, however, simply alternatives to one another: rather, they are complementary, one providing useful information where another fails. These differences arise through differences in wavelength and through differences in the interaction of the various types of radiation with matter; some of them will be considered in this chapter.

Firstly, we note that the wavelengths of light radiation are very much greater than interatomic spacings, and so diffraction maxima of the type observed with X-rays are not observed with light. This is not to say that light is not scattered by individual atoms: it is, but the only strong "diffracted" waves produced are the reflected and forward scattered waves, which latter determine the refractive index of the material. What are commonly classed as diffraction phenomena are only observed with apertures, or scattering bodies containing many thousands of atoms. Incidentally, much of the theory of diffraction of light is not applicable to electromagnetic radiation of much greater wavelength, e.g. radio- or microwaves, due to failure of St. Venant's principle: the wavelengths in these cases are comparable to the dimensions of diffracting bodies and boundary conditions become of great importance. We will not pursue this subject, but turn to a discussion of the interactions of X-rays, neutrons and electrons with atoms.

5.2. Scattering of X-rays

5.2a. *Introduction*

X-rays can interact with matter in a variety of ways. A satisfactory explanation of the phenomena can only be given by a wave-mechanical treatment and will not be attempted. Instead we shall outline the development of the classical theory, omitting a detailed derivation of the results. The classical theory yields results that are closely in agreement with the predictions of wave-mechanics and brings out clearly some of the physical principles involved. Its inadequacies are mostly conceptual and some of them will be indicated.

5.2b. *The classical theory*

On classical theory an electromagnetic wave incident on a free electron will cause it to oscillate about its mean position with a frequency equal to that of the incident radiation and in a direction parallel to the electric field component of the radiation, which is assumed to be plane polarised. Such an oscillating electric charge, since it is continuously being accelerated, will radiate at the frequency of the incident radiation. The scattered wave will have the same angular distribution of amplitude as that from an oscillating dipole, i.e. it will vary as $\sin \chi$, where χ is the angle between the dipole axis and the direction of observation. For an incident wave of unit amplitude, the scattered wave at a distance R will be given by

$$\phi = -\frac{1}{R}\frac{e^2}{mc^2} \sin \chi \exp\left[2\pi i(kR - vt)\right],$$

e and m being the electronic charge and mass and c the velocity of light. Note that e^2/mc^2 has the dimensions of length. This expression is an approximation to the field and is valid if $R \gg \lambda$.

Since X-ray wavelengths are of the same order of magnitude as atomic sizes, the path differences between a point of observation and different electrons in the atom will lead to significant phase differences between the waves scattered by these electrons. The path difference between waves from any two fixed points in the atom will be proportional to $\sin \theta$, where 2θ is the angle of scattering (cf. Fig. 4.1 and eq. (4.1)). The phase difference between these waves will therefore be proportional to $\sin \theta/\lambda$, and so the amplitude of the resultant wave from all the electrons in the atom will be a function of $\sin \theta/\lambda$. We define a quantity f_x, the atomic scattering factor for X-rays, as the ratio of the amplitudes of the waves scattered by the atom and by a single electron. f_x is, of course, a pure number. At $\sin \theta/\lambda = 0$ the scattered waves from all the electrons will be in phase and so $f_x = Z$, the number of electrons in the atom. At larger

D

values of $\sin \theta/\lambda$, f_x will be less, the decrease being greater for a large atom than for a small one, just as the amplitude of the diffracted wave from a slit aperture decreases more rapidly with angle away from the maximum for a wide slit than for a narrow one.

Values of f_x have been tabulated in a number of places (e.g. Cullity, 1956). It is only possible to do this usefully because the major part of the electron distribution in an atom is insensitive to the surrounding atoms:

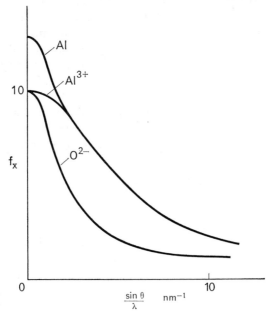

FIG. 5.1. Atomic scattering factors for X-rays of Al, Al^{3+} and O^{2-}.

only the outermost electrons take part in chemical bonding and are sensitive to neighbouring atoms, and these contribute very little to f_x at large values of $\sin \theta/\lambda$. At small values of $\sin \theta/\lambda$, however, the contribution of these electrons is not negligible, particularly in atoms of low atomic number, and so f_x curves for atoms and ions differ appreciably. This is illustrated for Al, Al^{3+} and O^{2-} in Fig. 5.1. These differences need to be borne in mind when calculating structure factors at small angles of scattering, particularly when the state of ionisation of an atom is not accurately known.

5.2c. *Quantum effects*

If the electrons in an atom really were free, as we have assumed, then the classical theory would certainly give incorrect results. The quantum aspects of electromagnetic radiation cannot be ignored when considering the interaction of electrons with high-energy X-ray quanta. A "billiard ball" treatment of a collision between a high-energy quantum and an electron predicts that the quantum will suffer a loss of energy, and hence an increase of wavelength, that depends on the angle through which it is scattered (the Compton effect). This is observed: in fact, Compton scattered radiation often constitutes a major part of the total scattered radiation. It does not, of course, give rise to diffraction effects since waves from different electrons will be incoherent, i.e. there is no constant phase relation between them. The reason that there is any coherent scattered radiation at all is that the electrons in an atom are not free but are bound to the atom, the atom as a whole absorbing any recoil. According to classical theory a bound electron would have a natural frequency of vibration, ω_0, depending on its mass and the strength of the binding. The results for a free electron would be valid provided that $\omega_0 \ll \omega$, the frequency of the incident radiation.

A perfectly free electron can scatter an incident quantum but it cannot absorb it, since it would not be possible in a quantum annihilation process to satisfy both the laws of conservation of energy and momentum. For even a loosely bound electron, however, this is not the case, as the atom can absorb some of the energy of the quantum and its recoil allows the electron to gain more momentum than there was in the quantum originally. Absorption processes of this kind, with the production of photoelectrons, are very important and lead to mass absorption co-efficients that depend on wavelength and atomic number but are generally of the order of 100 cm^2/gm. These are large enough to affect substantially the intensity of reflection from large, perfect crystals, particularly the weaker reflections.

Although attenuation of an X-ray beam by primary extinction (Section 4.5) can be much more rapid than the attenuation due to photoelectric absorption, photoelectric absorption is nevertheless the most significant source of attenuation in practice since it is effective in the whole volume of a specimen and not only in that very small fraction of it which is oriented at the correct angle to diffract strongly. Photoelectric absorption effectively limits the volume of a specimen that can be used in X-ray diffraction studies to a much lower value than is commonly used in neutron diffraction.

5.2d. *Anomalous scattering*

If the frequency of the incident radiation is comparable to or less than the natural frequency of vibration of a bound electron (which we identify with the frequency of the corresponding absorption edge), then the electron cannot be considered to be free. Both the amplitude and phase of the scattered wave become frequency dependent. At very low relative frequencies the amplitude of the scattered wave is negligible: classical theory therefore leads us to expect that the atomic scattering factor would be reduced by exactly two, the number of electrons in the K-shell, for wavelengths greatly in excess of that of the K absorption edge. Experimentally it is found that the reduction is not as great as this, varying with atomic number and being about 1·3 for elements of moderate atomic number. For wavelengths close to the absorption edge the reduction is greater. These, and similar effects associated with electrons in the L-shell, are adequately explained by the wave-mechanical theory. For a full discussion see James (1948).

Anomalous scattering, depending as it does on the relation of the wavelength of the incident radiation to the wavelengths of the absorption edges of the elements irradiated, can be utilised to distinguish between atoms of similar atomic number, which otherwise would have very similar atomic scattering factors and be difficult to distinguish with X-rays.

5.3. Scattering of neutrons

5.3a. *Introduction*

X-rays, as we have seen, are scattered by the electrons in an atom but not to any extent by the nuclei. With neutrons the reverse is generally true, and this is the cause of most of the differences between X-ray and neutron diffraction. In Sections 5.3 b–e we will consider some aspects of the scattering of neutrons by nuclei. Neutrons do also interact, through their magnetic moment, with any electrons with unpaired magnetic moments, and this will be considered in Section 5.3f.

5.3b. *The coherent scattering length*

Since a neutron is uncharged it does not interact electrostatically with a nucleus. The interaction is through the nuclear forces only and so different isotopes of the same element can (and sometimes do) have quite different scattering powers. Furthermore, since the neutron has a magnetic moment it can interact in two distinct ways with any nucleus that also possesses a magnetic moment, corresponding to parallel and antiparallel spin arrangements of the neutron and nucleus, and this can give rise to two quite different scattering powers even for the same nucleus. It follows that a crystal of, say, silver is not an assemblage of identical scattering centres but rather of four (spin parallel and spin antiparallel for each of the two isotopes) types of centres with different scattering powers, randomly distributed over the lattice sites. This leads to a diffuse background of scattered neutrons in addition to the strong Bragg reflections. The intensity of the latter is determined by the mean scattering power of the element, averaged over all isotopes and spin orientations, and it is this mean value, known as the coherent scattering length and denoted by b in neutron diffraction literature, that is the analogue of the atomic scattering power, $(e^2/mc^2) f_x$, in X-ray diffraction.

5.3c. *Neutron wavelengths and the variation of b with angle*

Neutron wavelengths are determined by the relation $\lambda = h/p$, where h is Planck's constant and p is the momentum of the neutron. Neutrons from a pile reactor will have a distribution of energies and hence of momenta and wavelengths: it is fortunate that the peak of the energy distribution corresponds to a wavelength of about 1 Å, since this is a convenient wavelength for investigating atomic structures. Since this wavelength is vastly greater than nuclear sizes, a nucleus acts essentially as a point scattering centre and so the scattered wave is spherically symmetrical, i.e. b is a constant, independent of $\sin \theta/\lambda$, unlike f_x. In consequence the intensity of high-order neutron diffraction peaks is greater in comparison with the intensity of low-order peaks than is the case with X-rays, though thermal vibrations of the atoms lead to a progressive decrease of intensity with increasing angle of diffraction in both cases.

5.3d. *The magnitude of b*

The coherent scattering length tends to increase slowly with increasing atomic number, Z, but there are large irregular variations superimposed, so neighbouring elements in the periodic table sometimes have very different scattering lengths. This can be of great value in determining the positions of elements of similar atomic number in a structure, since their X-ray scattering powers would be nearly the same and they would therefore be virtually indistinguishable by X-ray diffraction. The value of b is generally about 10^{-12} cm, so b is comparable with, though usually rather less than, X-ray scattering powers for elements of low Z. For elements of high Z the difference is greater, due to the steady rise of f_x with Z; for $Z \geqslant 29$ (Cu), $(e^2/mc^2)f_x$ at $\sin \theta/\lambda = 0$ is more than 10 times b; at $\sin \theta/\lambda = 0.5$ the same is true for $Z \geqslant 47$ (Ag), with a few exceptions among the rare earths. These values mean that dynamical effects become important only with rather larger crystals for neutron diffraction than for X-ray diffraction, particularly if heavy atoms are involved. The difference is not great, however, so for a crystal in which extinction is significant with X-rays it would probably also be significant with neutrons. In practice, secondary extinction is more important with neutrons than with X-rays on account of the fact that larger specimens are used (Section 5.3e) and that the effective volume (or penetration depth) is not reduced by absorption to the point where attenuation by secondary extinction can be neglected.

5.3e. *Absorption*

Absorption coefficients for neutrons are, with a few exceptions, very much less than they are for X-rays, typical values being 0.2 cm^{-1} and 1000 cm^{-1} respectively, though there is considerable variation between different elements. The small value of the absorption coefficient is of great practical importance as it means that large specimens may be used in neutron diffraction studies, thus counterbalancing relatively weak sources and low scattering powers. As far as the theory of diffraction is concerned it is of less importance, since absorption is normally only taken into account when calculating the effective volume of a specimen. It does, however, affect the shape of the intensity/angle curve in situations where dynamical interaction is important.

5.3f. *Magnetic scattering*

If an atom or ion in a crystal possesses a magnetic moment this can interact with the magnetic moment of a neutron and result in scattering. Magnetic scattering is considerably more complex than the scattering of X-rays by electrons, since it is a vector interaction depending on the orientation of the moments of the ion and neutron and also their relation to the Bragg reflecting planes. For a full discussion of this, and other aspects of the scattering of neutrons, see Bacon (1955). Here we will simply note the following points:

A. In a paramagnetic substance

  (i) the ion spins are randomly oriented and result only in incoherent scattering;

 (ii) this incoherent scattering depends on $\sin \theta/\lambda$ in much the same way, and for the same reasons, as does f_x. However, since the magnetic electrons are always in an outer shell of the ion, the decrease with increase in $\sin \theta/\lambda$ is more abrupt for magnetic scattering than for f_x;

(iii) the magnetic scattering length is of the same order of magnitude (10^{-12} cm) as the nuclear scattering length, b.

B. In ferromagnetic and antiferromagnetic substances, in which the ion spins are aligned,

  (i) scattered neutrons from different ions are coherent, and so give rise to diffraction maxima;

 (ii) the magnitude of the scattering in this case depends not only on $\sin \theta/\lambda$ but also on the orientation of the ionic spins relative to the Bragg reflecting planes. It is a maximum if the ion spins are in the planes and zero if they are normal to them;

(iii) with unpolarised neutrons the nuclear and magnetic scatterings are incoherent, and so the *intensities* of the nuclear and magnetic diffraction peaks are additive. Consequently the magnetic scattering can be considered entirely on its own, the nuclear scattering peaks being subtracted from the total scattering as though they were a part of the background;

(iv) with polarised neutrons the nuclear and magnetic scatterings are coherent: their relative phases depend on the orientation of the neutron spins and so the resultant *amplitude* of a diffraction peak may be the sum or difference of the nuclear and magnetic scattering amplitudes (or something intermediate). This can lead to a great increase in the sensitivity of detection of magnetic scattering.

5.4. Scattering of electrons

Electrons, being charged particles and very light, are strongly scattered by atoms. For a plane wave represented by

$$\Psi = \Psi_0 \exp[2\pi i(kz - vt)]$$

incident on an atom, the scattered wave can be represented by

$$\Psi = \Psi_0 \frac{f_e}{R} \exp[2\pi i(kR - vt)]$$

where f_e is the atomic scattering factor for electrons and has the dimensions of length, and the other symbols have their usual meanings. Since electron wavelengths are very small in comparison with atomic dimensions ($\lambda = 0.037$ Å for electrons of energy 100 keV), f_e varies rapidly with angle of scattering. It can be shown (see Hirsch *et al.*, 1965) that f_e is approximately given by

$$f_e = \frac{me^2}{2h^2} \left(\frac{\lambda}{\sin\theta}\right)^2 (Z - f_x). \tag{5.1}$$

The approximation is valid for high-energy electrons and atoms of moderate atomic weight. This expression is satisfactory for calculating values of f_e at angles such that $\sin\theta/\lambda \gtrsim 0.3$ Å^{-1}, but at small values of $\sin\theta/\lambda$, $f_x \to Z$ and small uncertainties in f_x lead to large uncertainties in f_e. Unfortunately the difference between Z and f_x at small values of $\sin\theta/\lambda$ is due almost entirely to interference between waves scattered by the outermost electrons in an atom and these are the most affected by adjacent atoms in a crystal.

At large values of $\sin\theta/\lambda$, $f_x \ll Z$ and so $f_e \propto Z$. It is not obvious from eq. (5.1) how f_e will vary with Z at small values of $\sin\theta/\lambda$. In fact it tends to increase, but not as rapidly, and perhaps not as consistently, as does f_x. This is illustrated for Al, Fe, Mo, Cs and Pb in Fig. 5.2.

Of more significance is the magnitude of f_e, which is mostly in the range 1–10 Å. This is very much greater than the scattering lengths for neutrons or X-rays which are mostly around 10^{-3} Å. Consequently dynamical effects are significant for very much smaller crystals in electron diffraction than in X-ray or neutron diffraction; indeed, they can hardly ever be ignored.

Electrons are not absorbed in the strict sense of that word, but they can be scattered inelastically with a consequent reduction in the amplitude of the incident electron wave that is equivalent to absorption. The corresponding "absorption" coefficient is much higher than for X-rays, but not in relation to the scattering length. A single absorption coefficient, however, is not adequate to describe the effects of absorption in the transmission of electrons through thin foils, for which anomalous absorption effects are found to be very important.

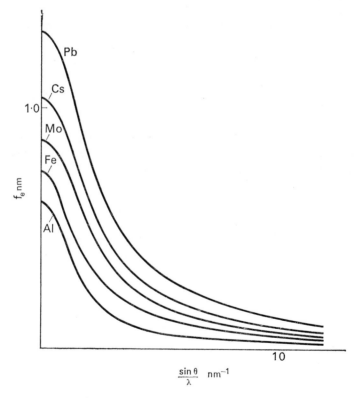

FIG. 5.2. Atomic scattering factors for electrons of Al, Fe, Mo, Cs and Pb.

D*

CHAPTER 6

FOURIER METHODS

6.1. Introduction

It was shown in Chapter 2 that a monochromatic beam of light incident on a sinusoidal grating would give rise to two and only two diffracted beams, one on either side of the incident beam. This result, together with the principle of superposition, ensures that Fourier analysis has an application to diffraction theory, for any grating can be considered to be a superposition of a possibly infinite number of sinusoidal gratings of different spacings, each of which will give rise to diffracted beams at different angles, and the amplitude of the beam at any angle will be proportional to the amplitude of the corresponding sinusoidal grating. In the following sections we will show how to determine these amplitudes, both for periodic and for non-periodic gratings. In many cases the calculation is no simpler than calculation by the methods of Chapter 2, but for complex gratings application of the convolution theorem, discussed in Section 6.5, is much simpler.

6.2. Trigonometrical series

6.2a. *Introduction*

Consider a periodic function $f(x)$, with period d. It will satisfy the relation

$$f(x+d) = f(x). \tag{6.1}$$

This may represent, for instance, the amplitude of a monochromatic wave after passing through a diffraction grating. It can be shown that, with the exception of some freak functions that do not occur in physics, any function of this form can be expressed as the sum of two infinite series, one of sine and one of cosine terms, plus a constant, thus

$$f(x) = \tfrac{1}{2}A_0 + \sum_{n=1}^{\infty} A_n \cos (2\pi nx/d) + \sum_{n=1}^{\infty} B_n \sin (2\pi nx/d). \tag{6.2}$$

(The reason for writing the constant term as $\frac{1}{2}A_0$ rather than as A_0 will appear later.)

The series are known as Fourier series, and this method of representing a periodic function is known as Fourier analysis.

6.2b. *Evaluation of the coefficients*

If it is possible to represent $f(x)$ in this form, which we shall assume to be the case, then the evaluation of the coefficients A_n and B_n is straightforward: for, multiplying each side of eq. (6.2) by $\cos{(2\pi mx/d)}$ and integrating over a complete period, which we may take as extending from $-\frac{1}{2}d \leqslant x < \frac{1}{2}d$, we have

$$\int_{-\frac{1}{2}d}^{\frac{1}{2}d} f(x) \cos{(2\pi mx/d)}\, dx = \int_{-\frac{1}{2}d}^{\frac{1}{2}d} \tfrac{1}{2}A_0 \cos{(2\pi mx/d)}\, dx$$

$$+ \int_{-\frac{1}{2}d}^{\frac{1}{2}d} \sum_{n=1}^{\infty} A_n \cos{(2\pi nx/d)} \cos{(2\pi mx/d)}\, dx$$

$$+ \int_{-\frac{1}{2}d}^{\frac{1}{2}d} \sum_{n=1}^{\infty} B_n \sin{(2\pi nx/d)} \cos{(2\pi mx/d)}\, dx$$

$$= \tfrac{1}{2}A_0 \int_{-\frac{1}{2}d}^{\frac{1}{2}d} \cos{(2\pi mx/d)} + \sum_{n=1}^{\infty} A_n \int_{-\frac{1}{2}d}^{\frac{1}{2}d} \cos{(2\pi nx/d)} \cos{(2\pi mx/d)}\, dx$$

$$+ \sum_{n=1}^{\infty} B_n \int_{-\frac{1}{2}d}^{\frac{1}{2}d} \sin{(2\pi nx/d)} \cos{(2\pi mx/d)}\, dx. \tag{6.3}$$

(We shall assume that term-by-term integration is permissible. It will be for all the functions that we are likely to encounter.)

Now

$$\sin{(2\pi nx/d)} \cos{(2\pi mx/d)} = \tfrac{1}{2}\{\sin{(2\pi(n+m)x/d)} + \sin{(2\pi(n-m)x/d)}\}, \tag{6.4}$$

$$\therefore \quad \int_{-\frac{1}{2}d}^{\frac{1}{2}d} \sin{(2\pi nx/d)} \cos{(2\pi mx/d)}\, dx = 0$$

since each sine term on the R.H.S. of eq. (6.4) is integrated over an integral number of periods. Likewise

$$\cos{(2\pi nx/d)} \cos{(2\pi mx/d)} = \tfrac{1}{2}\{\cos{(2\pi(n+m)x/d)} + \cos{(2\pi(n-m)x/d)}\},$$

$$\therefore \quad \int_{-\frac{1}{2}d}^{\frac{1}{2}d} \cos{(2\pi nx/d)} \cos{(2\pi mx/d)}\, dx = 0 \quad \text{if} \quad n \neq m,$$

$$= \tfrac{1}{2}d \quad \text{if} \quad n = m.$$

(The case $n = -m$ is excluded since n and m are both positive.)

$$\therefore \quad \int_{-\frac{1}{2}d}^{\frac{1}{2}d} f(x) \cos{(2\pi mx/d)}\, dx = \tfrac{1}{2}dA_m,$$

since the term with $n = m$ in the cosine series of eq. (6.3) is the only non-zero term.

$$\therefore \quad A_m = \frac{2}{d} \int_{-\frac{1}{2}d}^{\frac{1}{2}d} f(x) \cos(2\pi mx/d) \, dx.$$

Similarly,

$$B_m = \frac{2}{d} \int_{-\frac{1}{2}d}^{\frac{1}{2}d} f(x) \sin(2\pi mx/d) \, dx$$

and

$$A_0 = \frac{2}{d} \int_{-\frac{1}{2}d}^{\frac{1}{2}d} f(x) \, dx.$$

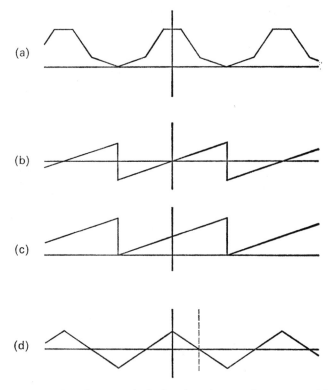

FIG. 6.1. Examples of some periodic functions that may be represented by sine or cosine series alone: (a) cosine series; (b) sine series; (c) sine series plus a constant; (d) either cosine or sine series.

6.2c. Cosine series

Very often the function $f(x)$ will possess some symmetry: if it is possible to choose an origin such that $f(x) = f(-x)$, as in Fig. 6.1a, then $B_m = 0$ for all m and the function can be represented by a cosine series alone, plus a constant term. Such a function is called an even function.

6.2d. Sine series

If it is possible to choose an origin such that $f(x) = -f(-x)$, as in Fig. 6.1b, then $A_m = 0$ for all m (including A_0), and the function can be represented by a sine series alone. Such a function is called an odd function. If the mean value of $f(x)$ is not zero, then this will not be possible, but it may be possible to convert the function into an odd function by subtracting a constant, i.e. it may be possible to find a constant c such that

$$f(x) - c = -\{f(-x) - c\},$$

as in Fig. 6.1c. In this case the function can be represented by a sine series plus a constant.

With some functions, such as that shown in Fig. 6.1d, it is possible to represent the function by either a sine or a cosine series, of course with different choices of origin in the two cases.

Single series representation is such a useful simplification that it is worth examining a function carefully to see if it is possible.

6.2e. Example

As an example of the determination of the coefficients we will evaluate the coefficients for the top-hat function shown in Fig. 6.2. The function is even if we choose an origin in the middle of a non-zero region, when it is defined as

$$f(x) = 0, \; -\tfrac{1}{2}d \leqslant x < -\tfrac{1}{2}a$$

$$f(x) = 1, \; -\tfrac{1}{2}a \leqslant x < \tfrac{1}{2}a$$

$$f(x) = 0, \; \tfrac{1}{2}a \leqslant x < \tfrac{1}{2}d$$

$$f(x+d) = f(x).$$

Since the function is even, $B_m = 0$.
Now

$$\int_{-\frac{1}{2}d}^{\frac{1}{2}d} = \int_{-\frac{1}{2}d}^{-\frac{1}{2}a} + \int_{-\frac{1}{2}a}^{\frac{1}{2}a} + \int_{\frac{1}{2}a}^{\frac{1}{2}d} = \int_{-\frac{1}{2}a}^{\frac{1}{2}a}$$

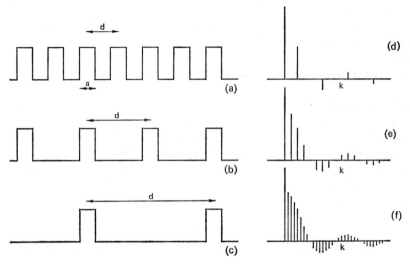

Fig. 6.2. Functions consisting of rectangular pulses of the same width, a, repeated at different distances, d, together with the coefficients of terms in the Fourier representations of these functions plotted against $k = m/d$.

for the integrals with which we shall be concerned, since the integrands are zero in the other intervals.

$$\therefore \qquad A_0 = \frac{2}{d} \int_{-\frac{1}{2}a}^{\frac{1}{2}a} dx = 2a/d.$$

Also

$$A_m = \frac{2}{d} \int_{-\frac{1}{2}a}^{\frac{1}{2}a} \cos\left(2\pi mx/d\right) dx$$

$$= \frac{2}{d} \frac{d}{2\pi m} \left[\sin\left(2\pi mx/d\right)\right]_{-\frac{1}{2}a}^{\frac{1}{2}a}$$

$$= \frac{2}{\pi m} \sin\left(\pi ma/d\right). \qquad (6.5)$$

For $d = 2a$,

$$A_0 = 1,$$

$$A_m = \frac{2}{\pi m} \quad \text{if} \quad m = 4l+1, \quad l = 0, 1, 2, \ldots$$

$$= \frac{-2}{\pi m} \quad \text{if} \quad m = 4l-1 \quad \Bigg\} \quad l = 1, 2, \ldots,$$
$$= 0 \quad \text{if} \quad m = 2l$$

$$\therefore \quad f(x) = \tfrac{1}{2} + \frac{2}{\pi} \{ \cos(2\pi x/d) - \tfrac{1}{3} \cos 3(2\pi x/d) + \tfrac{1}{5} \cos 5(2\pi x/d) - \ldots \}.$$

6.3. Exponential series

The periodic function $f(x)$ can also be represented by an infinite series of complex exponentials, thus

$$f(x) = \sum_{n=-\infty}^{\infty} C_n \exp(2\pi i n x/d). \tag{6.6}$$

The coefficients C_n are related to the coefficients of the sine and cosine series. Since

$$A_n \cos(2\pi n x/d) = \tfrac{1}{2} A_n \left(\exp(2\pi i n x/d) + \exp(-2\pi i n x/d) \right)$$

and

$$B_n \sin(2\pi n x/d) = \frac{B_n}{2i} \left(\exp(2\pi i n x/d) - \exp(-2\pi i n x/d) \right),$$

$$C_n = \frac{A_n}{2} + \frac{B_n}{2i} = \tfrac{1}{2}(A_n - i B_n),$$

also

$$C_{-n} = \frac{A_n}{2} - \frac{B_n}{2i} = \tfrac{1}{2}(A_n + i B_n)$$

and

$$C_0 = \tfrac{1}{2} A_0.$$

We note that C_n and C_{-n} are always complex conjugates if the function can be represented by eq. (6.2), i.e. if it is real. Furthermore, for an even function C_n and C_{-n} are real and equal, whereas for an odd function they are imaginary and opposites.

The coefficients C_n can also be obtained directly from eq. (6.6) by the methods used to obtain A_n and B_n, making use of the relations

$$\int_0^1 \exp(2\pi i m y) \exp(2\pi i n y) \, dy = 0 \quad \text{if} \quad m \neq -n,$$
$$= 1 \quad \text{if} \quad m = -n.$$

Multiplying both sides of eq. (6.6) by $\exp(-2\pi imx/d)$ and integrating we have

$$C_m = \frac{1}{d} \int_{-\frac{1}{2}d}^{\frac{1}{2}d} f(x) \exp(-2\pi imx/d) \, dx.$$

6.4. Fourier transforms

6.4a. *Analysis*

So far we have considered only functions that are periodic. No physical function will satisfy eq. (6.1) for all x since this, and also eqs. (6.2) and (6.6), represent functions of infinite extent, whereas a diffraction grating, even though it may consist of several thousand lines, is still finite. We will defer consideration of a finite grating until later and consider now how we can represent the function of Fig. 6.2a as d increases while a remains constant. It will be convenient for purposes of comparison at different values of d to define a new quantity $k_m = m/d$. k_m is, of course, the reciprocal of the repeat distance of the Fourier component of order m in the cosine representation of $f(x)$. It is defined only at multiples of the reciprocal distance d^{-1}. A plot of $A(k_m)$ against k_m is shown in Fig. 6.2d. For the function of Fig. 6.2a but with every alternate positive section removed, as in Fig. 6.2b, the plot would be as in Fig. 6.2e, while removing every alternate positive section from this function (Fig. 6.2c) would lead to the plot of Fig. 6.2f. The trend is clear: increasing the repeat distance reduces the interval of k_m between values at which $A(k_m)$ is defined but does not alter the form of the envelope of $A(k_m)$. Of course, the magnitude of each term is reduced, in proportion to the increase in d (or m, for the same value of k_m).

We might expect that, on proceeding to the limit of an infinite repeat distance, the values of k_m at which $A(k_m)$ is defined would become infinitely close together and $A(k_m)$ would become a continuous function of k_m. Now for this function,

$$A(k_m) = \frac{2}{\pi m} \sin(\pi a k_m)$$

$$= \frac{2a}{d} \frac{\sin(\pi a k_m)}{\pi a k_m},$$

so each value of $A(k_m)$ would go to zero, but the number of terms in a small interval of k_m would go to infinity and their sum would, we hope, remain finite.

For any even function, since $A(k_m)$ is defined at intervals of d^{-1}, the number of terms in a small interval δk_m ($\gg d^{-1}$) will be $\delta k_m d$ and their contribution to $f(x)$ will be

$$\delta(f(x)) = \delta k_m d \, A(k_m) \cos (2\pi k_m x)$$

provided that $A(k_m)$ does not change by much in the interval and that $\delta k_m x \ll 1$. If we let $\delta k_m \to 0$ and then proceed to the limit as $d \to \infty$ we have, dropping the subscript,

$$df(x) = d \, A(k) \cos (2\pi k x) \, dk,$$

$$\therefore \quad f(x) = 2 \int_0^\infty F(k) \cos (2\pi k x) \, dk$$

where

$$F(k) = \lim_{d \to \infty} d \, \frac{A(k)}{2}$$

$$= \int_{-\infty}^\infty f(x) \cos (2\pi k x) \, dx.$$

Since, if $f(x)$ is an even function $F(k)$ will also be an even function, we can obtain a pleasing symmetry between the relations for $f(x)$ and $F(k)$ by writing

$$f(x) = \int_{-\infty}^\infty F(k) \cos (2\pi k x) \, dk,$$

leaving unanswered what is meant by negative values of k.

In the case of a general periodic function requiring both sines and cosines for its representation, it is easier to use the complex form of the Fourier series, eq. (6.6). Then, proceeding as before, we find

$$F(k) = \int_{-\infty}^\infty f(x) \exp (-2\pi i k x) \, dx, \tag{6.7a}$$

$$f(x) = \int_{-\infty}^\infty F(k) \exp (2\pi i k x) \, dk. \tag{6.7b}$$

This carefree treatment of infinities and infinitesimals cannot be recommended to the mathematically minded. We will be content to accept the feasibility of transforming functions according to eq. (6.7) and leave it to the mathematicians to say under what circumstances it is not possible.

6.4b. *Example*

For the limiting function of Fig. 6.2, shown in Fig. 6.3a,

$$F(k) = \lim_{d \to \infty} d\,\frac{A(k)}{2}$$

$$= \frac{d}{2}\frac{2a}{d}\frac{\sin(\pi ak)}{(\pi ak)}$$

$$= a\,\frac{\sin(\pi ak)}{(\pi ak)}$$

This is also shown in Fig. 6.3a. Its square is, of course, the single slit diffraction pattern.

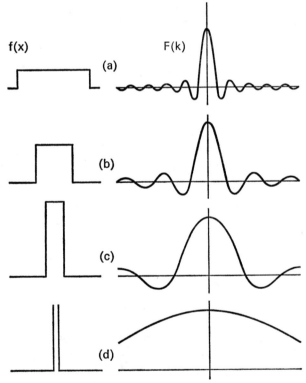

FIG. 6.3. Top-hat functions of different widths and their transforms.

The area under the function of Fig. 6.3a is just a. For a function of height h the area would be ha, and the transform $ha \sin(\pi ak)/(\pi ak)$. If we consider the sequence of functions shown in Fig. 6.3 we see that as $f(x)$ becomes narrower, $F(k)$ becomes wider. This is a general feature of transforms. In the limit as $a \to 0$, subject to the condition that $ha =$ constant, $F(k)$ approaches a constant $= ha$. The transform of other functions, for instance the Gaussian function $f(x) = \sigma^{-1} \exp(-x^2/\sigma^2)$, whose area is $\sqrt{\pi}$, also approaches a constant as the width approaches zero, as may easily be verified. It is obviously dangerous to try to infer what is the transform of the function $f(x) =$ constant.

6.5. Convolutions

6.5a. *Definition*

The convolution of two functions $f(x)$ and $g(x)$ is defined as

$$Q(y) = \int_{-\infty}^{\infty} f(x)g(y-x)\, dx.$$

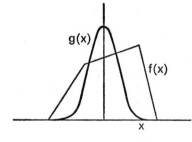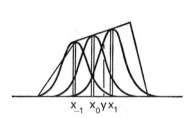

FIG. 6.4. The convolution, $Q(y)$, of two functions $f(x)$ and $g(x)$. The contributions to $Q(y)$ at any value of y from the small ranges δx of x indicated are the ordinates of the curves at that value of y, e.g. at y' the contributions are $f(x_{-1})\,\delta x \cdot g(y'-x_{-1}), f(x_0)\,\delta x \cdot g(y'-x_0)$ and $f(x_1)\,\delta x \cdot g(y'-x_1)$.

This is illustrated in Fig. 6.4. The contribution, $dQ(y)$, to $Q(y)$ from the range $x - \frac{1}{2}\delta x$ to $x + \frac{1}{2}\delta x$ of the variable x is obtained by setting down the function $g(y-x)$ (i.e. the function $g(y)$ with its origin at x) with a weighting factor of $f(x)\,\delta x$. $Q(y)$ is obtained by summing the contributions at $x = y$ of all curves such as those of Fig. 6.4, one for each range δx of x, and proceeding to the limit as $\delta x \to 0$. It does not matter which function we choose as f and which as g, the resulting convolution will be the same in each case, i.e.

$$Q(y) = \int_{-\infty}^{\infty} f(x)g(y-x)\,dx = \int_{-\infty}^{\infty} f(y-x)g(x)\,dx.$$

Usually, however, it is easier to visualise the convolution operation with one rather than the other of the two functions used as the weighting factor.

We have already encountered one example of a convolution: the intensity profile of a spectral line as seen in a spectrometer is the convolution of the "theoretical" line profile with the profile of the geometrical image of the spectrometer source. In this case it is most easily visualised if we suppose that each line element of the source is producing its own spectral line with an intensity proportional to the intensity of that part of the source, the spectral lines from different line elements in the source being displaced by amounts equal to the separation of the corresponding lines in the geometrical image of the source.

6.5b. *Convolutions and transforms*

Although there are numerous direct instances in physics of convolutions their greatest use is in connection with Fourier transforms. It can be shown that the transform of the product of two functions is the convolution of their individual transforms and, conversely, that the transform of the convolution of two functions is the product of their transforms. The proof is straightforward; for if $F(k)$ is the transform of $f(x)$ and $G(k)$ is the transform of $g(x)$, then the transform of $f(x)g(x)$ is

$$
\begin{aligned}
Q(k) &= \int f(x)g(x)\exp(-2\pi ikx)\,dx \\
&= \int \left(\int F(k')\exp[2\pi ik'x]\,dk'\right)g(x)\exp(-2\pi ikx)\,dx \\
&= \int F(k')\left(\int g(x)\exp[-2\pi i(k-k')x]\,dx\right)dk' \\
&= \int F(k')G(k-k')\,dk'.
\end{aligned}
$$

We will now use this theorem to discuss the transform of a finite diffraction grating.

6.5c. *Example*

An infinite grating with slits of width a separated by distances d between the centres of adjacent slits (Fig. 6.5c) is the convolution of an infinite grating of line sources (Fig. 6.5a) with the single-slit aperture function (Fig. 6.5b). The transform is then the product of the transforms of the component functions, as illustrated in Figs. 6.5 d–f.

A finite grating of the same form (Fig. 6.6c) is the product of an infinite grating (Fig. 6.6a) with the grating aperture function (Fig. 6.6b). The

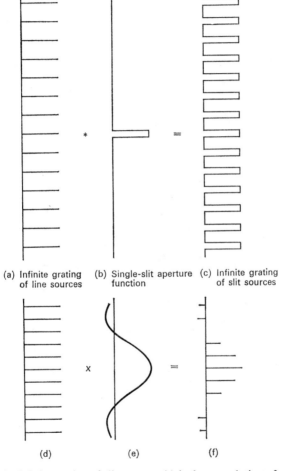

(a) Infinite grating   (b) Single-slit aperture   (c) Infinite grating
of line sources   function   of slit sources

(d)   (e)   (f)

FIG. 6.5. An infinite grating of slit sources (c) is the convolution of an infinite grating of line sources (a) with the single-slit aperture function (b). Its transform (f) is the product of the corresponding transforms (d and e).

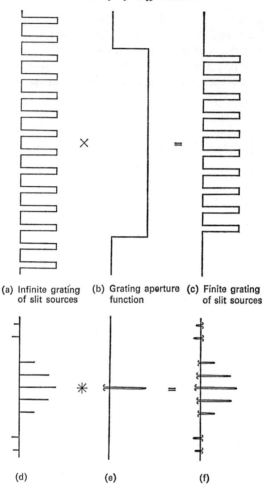

FIG. 6.6. A finite grating of slit sources (c) is the product of an infinite grating of slit sources (a) with the grating aperture function (b). Its transform (f) is the convolution of the corresponding transforms (d and e).

transform of the finite grating is then the convolution of the transforms of the component functions, as illustrated in Figs. 6.6 d–f.

Translating the variable k of the function of Fig. 6.6f into the value of $\sin \theta/\lambda$ at which a diffracted beam appears when the grating is illuminated at normal incidence with monochromatic light, and squaring the amplitude, we get the intensity versus angle curve for this grating given in Fig. 2.12.

6.6. Transforms in two and three dimensions

Consider a doubly periodic function m of x and y, with repeat distances d_1 and d_2. m may be, for instance, the mass per unit area of a lamina. If we consider a line parallel to the x-axis at a distance y from it, then the value of m at any point on this line is a periodic function of x and may be represented by a Fourier series

$$m(x, y) = \sum_{h=-\infty}^{\infty} M_h \exp(2\pi ihx/d_1). \tag{6.8}$$

A similar representation is possible for any other line, but the coefficients M_h will not in general be the same. They will be the same, however, for lines at $y+d_2$, $y+2d_2$, etc., since these are identical lines. M_h is therefore a periodic function of y, with period d_2, and so can be represented by the Fourier series

$$M_h = \sum_{k=-\infty}^{\infty} M_{hk} \exp(2\pi iky/d_2).$$

Substituting this in eq. (6.8),

$$m(x, y) = \sum_{h=-\infty}^{\infty} \sum_{k=-\infty}^{\infty} M_{hk} \exp[2\pi i(hx/d_1 + ky/d_2)].$$

Note that this argument does not presuppose that the axes of x and y are orthogonal.

To each and every pair of numbers h and k there corresponds one term $M_{hk} \exp[2\pi i(hx/d_1 + ky/d_2)]$ in the double series (though the coefficients may not all be different from zero). By a simple extension of this argument a triply periodic function, μ, may be represented by a triple series

$$\mu(x, y, z) = \sum_{h=-\infty}^{\infty} \sum_{k=-\infty}^{\infty} \sum_{l=-\infty}^{\infty} F_{hkl} \exp[2\pi i(hx/d_1 + ky/d_2 + lz/d_3)]. \tag{6.9}$$

If μ represents the distribution of scattering power within a crystal, then F_{hkl} is the structure factor for the *hkl* reflection, as we shall now proceed to show.

The notation is simplified if we write $x = ud_1$, $y = vd_2$ and $z = wd_3$.

Then, multiplying both sides of eq. (6.9) by $\exp\left[-2\pi i(h'u+k'v+l'w)\right]$ and integrating over the range $0 \leqslant u, v, w < 1$, we have

$$\int_0^1\int_0^1\int_0^1 \mu(u, v, w) \exp\left[-2\pi i(h'u+k'v+l'w)\right] du\,dv\,dw$$

$$= \sum_h \sum_k \sum_l F_{hkl}[\int_0^1 \exp\left[2\pi i(h-h')u\right] du \int_0^1 \exp\left[2\pi i(k-k')v\right] dv$$

$$\times \int_0^1 \exp\left[2\pi i(l-l')w\right] dw].$$

Now

$$\int_0^1 \exp\left[2\pi i(h-h')u\right] du = 0 \quad \text{if} \quad h \neq h',$$

$$= 1 \quad \text{if} \quad h = h'.$$

Therefore the only non-zero term in the triple summation is the term for which $h = h', k = k', l = l'$, and this term is just $F_{h'k'l'}$.

$$\therefore \quad F_{h'k'l'} = \int_0^1\int_0^1\int_0^1 \mu(u, v, w) \exp\left[-2\pi i(h'u+k'v+l'w)\right] du\,dv\,dw.$$

Comparing this with eq. (4.3), which was derived for a discrete distribution of scattering centres, we see that $\mu(u, v, w)\,du\,dv\,dw$, the scattering power of the matter within the element of volume at u, v, w, is equivalent to f and that $F_{h'k'l'}$, the coefficient of a term in the Fourier representation of the distribution of scattering power, is also the structure factor for the $h'k'l'$ reflection.

Fourier integrals and transforms can be developed in three dimensions just as in one, and the convolution theorem is also valid and useful. Some points that we may note without proof, as an example of the way in which Fourier analysis can be useful, are that the transform of a lattice is the reciprocal lattice, that an infinite crystal structure is the convolution of a unit cell with the lattice and that its transform is therefore the product of the transform of a unit cell (i.e. the structure factor) with the reciprocal lattice, leading to the weighted reciprocal lattice, and that a crystal is the product of a shape function with a crystal structure: its transform is therefore the convolution of the transform of the shape function with the weighted reciprocal lattice, which we have already encountered as the square root of the interference function.

CHAPTER 7

ELEMENTS OF CRYSTALLOGRAPHY

7.1. Introduction

A crystal is a three-dimensional array of atoms (or ions) in which, apart from minor irregularities, a group of atoms is repeated at regular intervals in space. It is because of this regularity that crystals develop planar faces at well-defined angles to one another and that the shape of a crystal is often beautifully symmetric. The early study of crystallography was devoted almost entirely to description and classification of the external forms of crystals. Indeed, materials with no well-developed faces were classified as non-crystalline. Nowadays the emphasis is on the internal structure and it is recognised that some materials without any external regularity of shape may nevertheless possess internal regularity and so should be classified as crystals. Not all solids are crystalline, however: glass is an example of a solid that does not possess the degree of internal regularity that is characteristic of crystals.

Crystals are classified and described according to the regularity of their atomic arrangement. It is convenient to think of the crystal as based on a regularly repeating array of points, a lattice, with an identical grouping of atoms around each lattice point. However, it is important to be clear of the distinction between the crystal structure and the mathematical abstraction, the lattice, on which it is based. In some very simple crystals, such as those of many metals, the repeating unit consists of just one atom; the lattice point is naturally taken to be at the centre of this atom. However, in a complex structure with a large number of atoms in each repeating group, there may be no obvious way of choosing the group or of lattice point within a group. The number of atoms within a group will always be the same, but the shape of the group and the position of the lattice point within it can be chosen arbitrarily. This is illustrated in Fig. 7.1, from which it can be seen that, regardless of the choice of group or of lattice point within a group, for a given structure the relation of one

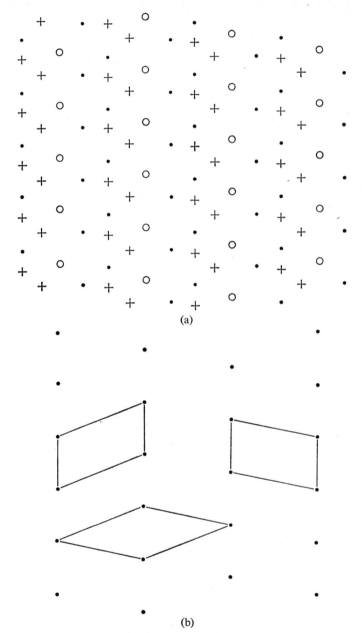

FIG. 7.1. (a) A two-dimensional structure. (b) The lattice on which the structure of Fig. 7.1a is based, showing three possible unit cells.

lattice point to another is always the same. It is because of this that the geometry of lattices is a suitable starting point for the study of crystallography.

The lattice alone does not tell us all there is to know about the regularity of a structure, since it tells us nothing about the regularity of the arrangement of atoms within a group. This is a much more complex subject, and one that we will not consider.

7.2. Lattices

7.2a. *Definition*

A *lattice* is an array of points such that the environment of any one point is identical with that of all others. Alternatively, it may be defined as the set of points at the ends of the vectors

$$\mathbf{r} = n_1\mathbf{a}_1 + n_2\mathbf{a}_2 + n_3\mathbf{a}_3$$

where n_1, n_2 and n_3 can take all integral values, positive, negative or zero, and $\mathbf{a}_1$, $\mathbf{a}_2$ and $\mathbf{a}_3$ are fixed, non-coplanar vectors, the *base vectors* of the lattice.

7.2b. *Unit cells*

The vector definition of a lattice leads at once to the concept of a *unit cell*. The unit cell is the parallelepiped whose edges are the vectors $\mathbf{a}_1$, $\mathbf{a}_2$ and $\mathbf{a}_3$. As is apparent from the two-dimensional case illustrated in Fig. 7.1b, the choice of unit cell and hence of base vectors to generate a given lattice is not unique: there is, in fact, an infinite number of possible unit cells for each lattice.

The lattice can be thought of as an assembly of unit cells, packed together corner to corner to fill all space, with a lattice point at each corner. We could equally well visualise the lattice as an assembly of larger cells, all identical and again fitting together corner to corner to fill all space, but now there will be lattice points at other positions as well as at the corners of the cells. The edge vectors of such a cell will, of course, only generate the lattice points at the corners of the cells; in order to specify the positions of all the lattice points we must also specify the positions of the additional lattice points within each cell. This can be done by means of one or more *basis vectors*, i.e. vectors from the origin of a unit cell to the positions of the additional lattice points associated with that cell.

Cells containing more than one lattice point per cell are still referred to as "unit" cells. They are unit cells in the sense that they are units from which the lattice can be built, although they are not unitary in the sense of containing only one lattice point. Unit cells containing only one lattice point are known as *primitive* unit cells: other cells are named according to the positions of the additional lattice points.

It might appear that it would always be simpler to describe a given lattice in terms of one of the many possible primitive unit cells. This is not so, since it is sometimes possible to find a non-primitive unit cell that is more symmetrical than the most symmetrical primitive unit cell; in particular, to choose a unit cell with base vectors that are orthogonal. The advantages of so doing easily outweigh the (slight) disadvantage of working with a non-primitive unit cell. In the following sections we will consider the symmetry of lattices, and see why this is so.

7.2c. *Symmetry*

Lattices can, and generally do, possess certain elements of symmetry. A *symmetry element* is defined as an operation which, when performed on the lattice, produces an identical lattice. All lattices possess translational symmetry: displacing the lattice parallel to itself by any of the vectors **r** leaves the lattice unaltered. They also possess centres of symmetry, or inversion symmetry: inverting the lattice through any of the lattice points, or indeed through some other points, will again produce an identical lattice. Of more interest are symmetry elements of rotation and reflection. A lattice is said to possess an *n-fold axis of rotation* if rotation of the lattice through an angle $2\pi/n$ about some fixed axis produces an identical lattice. (Another way of looking at it is that an observer in the lattice would get an identical view if he turned through an angle $2\pi/n$.) The lattice is said to possess a *mirror plane* if reflection of the lattice in some plane produces an identical lattice.

The symmetry of a crystal structure is not necessarily the same as the symmetry of the lattice on which it is based. Whether it is or not will depend on the arrangement of atoms around each lattice point. For instance, although every *lattice* possesses a centre of symmetry, the two-dimensional structure represented in Fig. 7.1a does not.

With crystals it is also possible to identify more complex symmetry operations, involving translations as well as reflections or rotations. This leads to a much greater variety of possible combinations of symmetry elements, or *space groups*, for a crystal structure than for a lattice. We will not pursue this subject but will confine our discussion to the symmetry of lattices.

7.2d. *Crystal systems*

Crystals (and lattices) are divided into seven crystal *systems*, according to the symmetry of the most symmetrical unit cell it is possible to find in the structure. The unit cell can be described either in terms of the symmetry elements it possesses, or alternatively in terms of the relations between its axes. If we let the lengths of the cell edges be a, b and c, and the angles between b and c, c and a, and a and b be α, β and γ, respectively, then in a lattice with no more than minimum symmetry we would expect to find $a \neq b \neq c$ and $\alpha \neq \beta \neq \gamma \neq 90°$. Such a lattice is assigned to the *triclinic* system. The seven crystal systems, with the relations between the axes in each, are shown in Table 7.1.

TABLE 7.1

System	Relations between axes		
Triclinic	$a \neq b \neq c$;	$\alpha \neq \beta \neq \gamma \neq 90°$	
Monoclinic	$a \neq b \neq c$;	$\alpha = \gamma = 90°$;	$\beta \neq 90°$
Orthorhombic	$a \neq b \neq c$;	$\alpha = \beta = \gamma = 90°$	
Trigonal	$a = b = c$;	$\alpha = \beta = \gamma \neq 90°$	
Tetragonal	$a = b \neq c$;	$\alpha = \beta = \gamma = 90°$	
Hexagonal	$a = b \neq c$;	$\alpha = \beta = 90°$;	$\gamma = 120°$
Cubic	$a = b = c$;	$\alpha = \beta = \gamma = 90°$	

The labelling of the axes, for instance in the monoclinic system where $\alpha = \gamma = 90°$ and $\beta \neq 90°$, is conventional.

One might well ask why there is no crystal system in which, for instance, $a = b = c$ but $\alpha \neq \beta \neq \gamma$. The reason is that the classification is according to symmetry and there is no combination of symmetry elements that would give rise to a unit cell with the above relations between its axes in which it was not possible to find a more symmetrical unit cell.

Since the symmetry of a structure is not necessarily as great as the symmetry of the lattice on which it is based, and as a discussion of the symmetry of structures would be out of place here, we will not consider the minimum symmetry requirements of a crystal structure in each of the crystal systems. The symmetries of the lattices can easily be obtained by inspection of the unit cells except in the hexagonal system, which is a somewhat special case.

7.2e. *Bravais lattices*

As mentioned in Section 7.2c, it may be possible to find in a lattice a non-primitive unit cell with more symmetry than the most symmetrical primitive unit cell. For instance, if an extra lattice point were to be added at the centre of every (primitive) unit cell of an orthorhombic lattice, as in Fig. 7.2a, the symmetry of the resulting lattice would be the same as

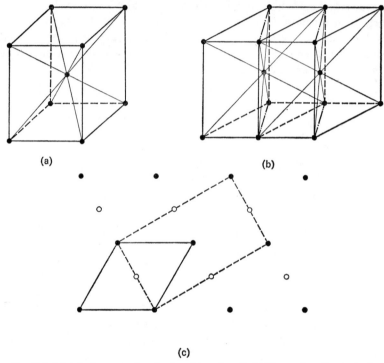

(a)   (b)

(c)

FIG. 7.2. (a) A body-centred orthorhombic unit cell. (b) Two unit cells of a body-centred monoclinic lattice, showing the alternative unit cell that is centred on one pair of faces. (c) Plan, projected on plane normal to six-fold axis, of a hexagonal lattice (full circles), to which extra lattice points (open circles) at heights $\frac{1}{2}c$ above the plane of projection have been added. There is no six-fold axis of symmetry in the composite lattice. A face-centred orthorhombic unit cell is outlined with the broken lines.

before, but it is no longer possible to find a primitive unit cell with this symmetry. The most symmetrical primitive unit cell would, in fact, have no more than the minimum symmetry possessed by every lattice (though it is possible to find one with $a = b = c$), and to describe the lattice in terms of this unit cell, suggesting that it belongs to the triclinic system, would be misleading. Obviously, however, the lattice is different from the primitive orthorhombic lattice with which we started (though not as regards its symmetry). We recognise it as a distinct lattice type, a *body-centred orthorhombic* lattice. Similarly we can recognise body-centred lattices in the tetragonal and cubic systems, but not in the other systems: in the triclinic and trigonal systems it is still possible to find a primitive unit cell with the full symmetry of the lattice; in the monoclinic system the lattice produced is distinct from the original lattice, but is equivalent to the lattice obtained by adding points to the centres of one pair of faces, as shown in Fig. 7.2b, and this is the conventional description of this lattice; while in the hexagonal system adding a lattice point to the centre of each unit cell alters the symmetry of the lattice: the six-fold axis of rotation is reduced to one of two-fold symmetry and the lattice is equivalent to an orthorhombic lattice with additional lattice points at the centres of all faces. This is illustrated in the plan drawing of Fig. 7.2c.

Proceeding in this way we find that it is possible to distinguish fourteen distinct lattice types. These are known as Bravais lattices, and are listed in Table 7.2.

It should be borne in mind that in any lattice it is always possible to choose any number of unit cells, either primitive or larger (though

TABLE 7.2. BRAVAIS LATTICES

Triclinic	P			
Monoclinic	P,	C		
Orthorhombic	P,	C,	F,	I
Trigonal	P			
Tetragonal	P,	I		
Hexagonal	P			
Cubic	P,	F,	I	

P   Primitive.
C   Centred on faces opposite to *c*-axis.
F   Centred on all faces.
I   Body-centred.

not, of course, all with the symmetry of the lattice), and it may be advantageous to refer a structure to one of these cells rather than to the Bravais cell. Even a very small distortion of a structure can destroy a symmetry element and so the similarity between two crystal structures based on different Bravais lattices may be obscured if the Bravais cells are used. For example, the rhombohedral structure of mercury, illustrated in Fig. 7.3, is not very different from the face-centred cubic structure of many other metals.

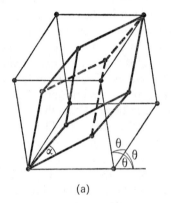
(a)

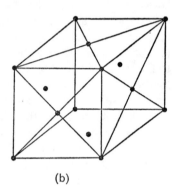
(b)

FIG. 7.3. (a) Part of the structure of mercury. The trigonal unit cell, with interaxial angles $\alpha = 70°32'$, is outlined with heavy lines. The angle $\theta = 98°10'$. (b) The unit cell of a face-centred cubic lattice. The angle corresponding to α in this structure equals 60°.

7.3. Indices

7.3a. *Introduction*

Lattice points can be thought of as lying on sets of parallel lines or sets of parallel planes, as illustrated for two dimensions in Fig. 7.4. The only sets of planes that are of importance in diffraction are those on which lie all the lattice points. However, if a lattice is referred to a unit cell with a basis, then it is convenient to disregard for the time being those points that are not at the corners of the unit cells. In the following subsections on indices any reference to lattice points will mean only the corner points. Of course, if the lattice is referred to a primitive unit cell, this will be all the lattice points.

We need some method of referring to a particular set of lines or planes. The method universally adopted is that of Miller indices, which will now be described.

7.3b. *Indices of a set of lines*

The indices of a set of lines are defined as follows: consider that line of the set that passes through the origin. If the nearest lattice point to the origin on the line is the point at $U\mathbf{a}_1 + V\mathbf{a}_2 + W\mathbf{a}_3$ with reference to a

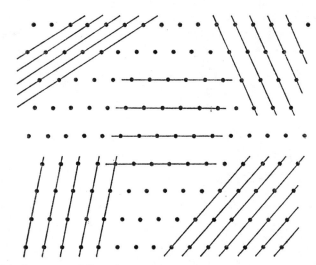

FIG. 7.4. Some sets of lines that contain all the points of a two-dimensional lattice.

particular set of base vectors, then $[UVW]$ are the indices of the line. If U, V or W is negative this is denoted by placing a bar over the symbol, thus, $\overline{U}$, and is read "bar U".

The square brackets, [ ], indicate that the indices refer to a set of lines or a direction. In any lattice possessing symmetry there will be a number of other directions related to the first by the symmetry elements of the lattice. If the unit cell chosen has the symmetry of the lattice (which, remember, is not possible in the hexagonal system), then the indices of these symmetry-related directions will be permutations of $\pm U$, $\pm V$ and $\pm W$. Not all possible permutations need be symmetry-related directions,

E

of course, but there will be no symmetry-related directions that are not permutations of these few symbols. This is part of the reason for choosing a unit cell with the lattice symmetry. The set of symmetry-related directions is denoted by $\langle UVW \rangle$, with angle brackets.

7.3c. *Indices of a set of planes*

Consider that plane of the set that passes through the origin: if the adjacent plane makes intercepts of $\mathbf{a}_1/h$, $\mathbf{a}_2/k$ and $\mathbf{a}_3/l$ on the three axes, then hkl are the indices of the set. If the set of planes is parallel to one or more of the axes, then the corresponding intercepts are infinite and the indices are zero. Indices of a few sets of planes are illustrated in Fig. 7.5.

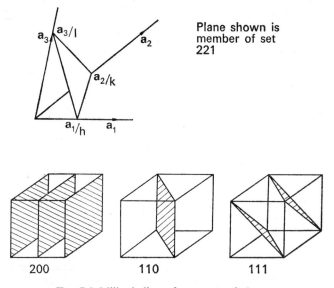

FIG. 7.5. Miller indices of some sets of planes.

Note that fractional indices are not possible for crystallographically important sets of planes, i.e. sets of planes that contain all the lattice points: for if a plane made a fractional intercept of, say, 2/5 on one of the axes, the next planes would make intercepts of 4/5 and 6/5 and the nearest lattice point to the origin along this axis would not lie on any plane of the set.

In the early days of crystallography interest was centred on the faces developed by crystals. For consideration of these all that mattered was the orientation of a set of planes, not their spacing, and so no distinction was made between what we have defined as, say, the 110 and 220 sets of planes. Consequently, in the interests of simplicity, all sets of indices were cleared of any common factors. A set of indices enclosed in parentheses, e.g. (*hkl*), denotes a crystal face parallel to the *hkl* set of planes. In this representation *h*, *k* and *l* should contain no common factor.

Crystal faces and sets of planes are also related by symmetry. The symbol {*hkl*}, with curly brackets, denotes the set of faces that are related by symmetry to the face (*hkl*). The same representation is used to denote related sets of planes; in this case *h*, *k* and *l* may contain a common factor.

7.3d. *The hexagonal system*

Because it is not possible to choose a unit cell with hexagonal symmetry, crystals in the hexagonal system are generally indexed on a slightly different system, known as Miller–Bravais indices. A fourth index, *i*, is added to the indices of a plane, which are written (*hkil*). *i* is defined by the relation

$$h+k+i = 0.$$

It may be thought of as the reciprocal of the fractional intercept on a fourth axis, commonly called *u*, which is coplanar with the *x*- and *y*-axes and symmetrically related to them, as in Fig. 7.6.

The advantage of Miller–Bravais indices can be seen from the following example: the Miller indices of the set of faces related by symmetry to the face (110) are (110); ($\bar{1}$20); ($\bar{2}$10); ($\bar{1}\bar{1}$0); (1$\bar{2}$0); (2$\bar{1}$0). The fact that these faces are crystallographically identical is not very obvious from their indices. The Miller–Bravais indices of the same faces are (11$\bar{2}$0); ($\bar{1}$2$\bar{1}$0); ($\bar{2}$110); ($\bar{1}\bar{1}$20); (1$\bar{2}$10); (2$\bar{1}\bar{1}$0), from which the relation is much more obvious.

Miller–Bravais indices can also be used for directions, but not as simply. They are not obtained by adding a fourth index to an existing three, as with planes, but by finding vectors along the four axes whose vector sum is in the required direction and such that the algebraic sum of the first three indices is zero. The direction [100], for instance, becomes [2$\bar{1}\bar{1}$0], and [230] becomes [14$\bar{5}$0]. The chief objection to this system is that the indices of a direction are sometimes difficult to discover or interpret: the chief advantage is that the direction [*hkil*] is somewhere near to the normal to the face (*hkil*)—in special cases, parallel to it, as in the example of Fig. 7.6.

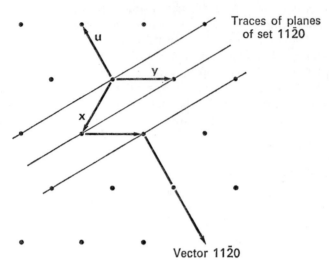

Fig. 7.6. Miller–Bravais indices in hexagonal system.

7.3e. *Interplanar spacings*

The spacing, d_{hkl}, between successive planes of the set *hkl* can be expressed in terms of the edge lengths *a*, *b* and *c* and interaxial angles α, β and γ of the unit cell. For the triclinic system the relation is

$$d = \frac{abc(1-\cos^2 \alpha - \cos^2 \beta - \cos^2 \gamma + 2 \cos \alpha \cos \beta \cos \gamma)^{\frac{1}{2}}}{(S_{11}h^2 + S_{22}k^2 + S_{33}l^2 + 2S_{12}hk + 2S_{23}kl + 2S_{13}hl)^{\frac{1}{2}}}$$

where

$$S_{11} = b^2 c^2 \sin^2 \alpha,$$
$$S_{22} = a^2 c^2 \sin^2 \beta,$$
$$S_{33} = a^2 b^2 \sin^2 \gamma,$$
$$S_{12} = abc^2(\cos \alpha \cos \beta - \cos \gamma),$$
$$S_{23} = a^2 bc(\cos \beta \cos \gamma - \cos \alpha),$$
$$S_{13} = ab^2 c(\cos \alpha \cos \gamma - \cos \beta).$$

In more symmetrical unit cells this expression can be simplified, using the relations between the axes given in Table 7.1. For instance, in the trigonal system

$$d = \frac{a(1 - 3\cos^2\alpha + 2\cos^3\alpha)^{\frac{1}{2}}}{(h^2 + k^2 + l^2)\sin^2\alpha + 2(hk + kl + hl)(\cos^2\alpha - \cos\alpha)^{\frac{1}{2}}}.$$

For the primitive unit cell shown in Fig. 7.7, $\alpha = 109° 28'$ ($\cos\alpha = -\frac{1}{3}$), and so

$$d = \frac{a(2/3)^{\frac{1}{2}}}{(h^2 + k^2 + l^2 + hk + kl + lh)^{\frac{1}{2}}}.$$

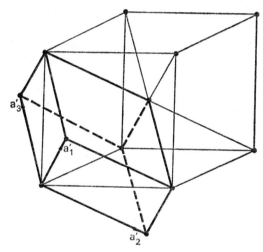

FIG. 7.7. Body-centred cubic lattice, with primitive and body-centred unit cells outlined. Cell edges of primitive unit cell are vectors $a_1' = \frac{1}{2}(-1, 1, 1)$, $a_2' = \frac{1}{2}(1, -1, 1)$ and $a_3' = \frac{1}{2}(1, 1, -1)$ with respect to axes of body-centred cell.

However, when indexed on the body-centred unit cell with cubic symmetry the relation is

$$d = \frac{a}{(h^2 + k^2 + l^2)^{\frac{1}{2}}}.$$

The advantage of working with the most symmetrical unit cell in the lattice should now be obvious.

Expressions for the interplanar spacings in all systems are given by Cullity (1956).

7.4. The reciprocal lattice

7.4a. *Definition*

Consider a lattice defined by the vectors a_1, a_2 and a_3. The lattice defined by the vectors b_1, b_2 and b_3, such that

$$a_i \cdot b_j = \delta_{ij}, \quad \text{where} \quad \begin{aligned} \delta_{ij} &= 1 \quad \text{if} \quad i = j \\ &= 0 \quad \text{if} \quad i \neq j \end{aligned} \right\} \tag{7.1}$$

stands in a special relation to the first lattice, and is known as the reciprocal lattice. Obviously, from the symmetry of the defining relation, the lattice that is reciprocal to the **b** lattice is the **a** lattice, so the two lattices form a reciprocally related pair (as is implied, but not proved, by the name). An equivalent, though more cumbersome, statement of the relation is that

$$b_1 = \frac{a_2 \wedge a_3}{a_1 \cdot a_2 \wedge a_3}, \quad b_2 = \frac{a_3 \wedge a_1}{a_2 \cdot a_3 \wedge a_1}, \quad b_3 = \frac{a_1 \wedge a_2}{a_3 \cdot a_1 \wedge a_2} \tag{7.2}$$

with similar relations for the **a** vectors. It is easily verified that the vectors defined by eq. (7.2) satisfy eq. (7.1).

7.4b. *Lattices with a basis*

If the vectors a_1, a_2 and a_3 are not a complete description of a lattice, i.e. they are merely the edge vectors of a unit cell with a basis, then the vectors b_1, b_2 and b_3, as defined by eq. (7.1), will not define the reciprocal lattice. As the **a** vectors by themselves define only a fraction of the lattice points in the **a** lattice, so the **b** vectors define more points than belong to the reciprocal lattice (the reciprocal relation being apparent in this). The true reciprocal lattice may be found by choosing a different set of base vectors in the **a** lattice, outlining a primitive cell (though this is not the usual procedure).

It should be emphasised that the lattice that is reciprocal to a given lattice exists quite independently of any description of either, i.e. of any choice of unit cell. The reciprocal lattice has been defined in terms of a particular set of base vectors, but whatever the choice of base vectors, the resulting reciprocal lattice (though not its description) will be the same.

7.4c. *Example*

As an example we will consider the lattice that is reciprocal to a face-centred cubic lattice. If we take axes along the edges of the conventional unit cell (i.e. the face-centred cell), and let the edges of this cell be of length a, then a suitable set of base vectors to define a primitive cell would be the vectors $\mathbf{a}_1 = (0, \frac{1}{2}a, \frac{1}{2}a)$, $\mathbf{a}_2 = (\frac{1}{2}a, 0, \frac{1}{2}a)$, $\mathbf{a}_3 = (\frac{1}{2}a, \frac{1}{2}a, 0)$, i.e. the vectors $\frac{1}{2}a(0, 1, 1)$, $\frac{1}{2}a(1, 0, 1)$, $\frac{1}{2}a(1, 1, 0)$. (These are components of a vector, not indices of a direction; the term outside each bracket is to be understood as multiplying each component of the vector inside the bracket.)

Now
$$\mathbf{a}_2 \wedge \mathbf{a}_3 = \tfrac{1}{4}a^2(-1, 1, 1)$$
and
$$\mathbf{a}_1 \cdot \mathbf{a}_2 \wedge \mathbf{a}_3 = \tfrac{1}{4}a^3,$$
$$\mathbf{b}_1 = \frac{1}{a}(-1, 1, 1) = \frac{2}{a}(-\tfrac{1}{2}, \tfrac{1}{2}, \tfrac{1}{2}).$$

Similarly,
$$\mathbf{b}_2 = \frac{1}{a}(1, -1, 1)$$
and
$$\mathbf{b}_3 = \frac{1}{a}(1, 1, -1).$$

These vectors are the vectors from the origin to three of the body-centring positions of a body-centred cubic lattice, of cell edge $2/a$, and outline a primitive unit cell in this lattice. This is illustrated in Fig. 7.7.

We have seen that the reciprocal lattice to a face-centred cubic lattice is a body-centred cubic lattice. It follows that the reciprocal of a body-centred cubic lattice is a face-centred cubic lattice. This is not the usual, nor the simplest, way of deriving these results. The usual and simplest way is by reference to the structure factor, as will be explained in the next section.

7.5. Relations between lattices

7.5a. *Introduction*

The reciprocal lattice is useful because of the relation it bears to the crystal lattice. We have already seen an example of this: in Section 4.2 it was found convenient to refer the wave vector of an incident beam of radiation to the base vectors of the reciprocal lattice, since this led to a simplification of the treatment and a geometrical interpretation of the result. In this section we will prove some of the relations between the direct and reciprocal lattices that are useful in diffraction theory.

7.5b. *The product relation*

If $\mathbf{g} = h\mathbf{b}_1 + k\mathbf{b}_2 + l\mathbf{b}_3$ is a vector of the reciprocal lattice and $\mathbf{r} = n_1\mathbf{a}_1 + n_2\mathbf{a}_2 + n_3\mathbf{a}_3$ a vector of the crystal lattice, then

$$\mathbf{g} \cdot \mathbf{r} = n_1 h + n_2 k + n_3 l,$$

an integer.

This result follows at once from the relation $\mathbf{a}_i \cdot \mathbf{b}_j = \delta_{ij}$.

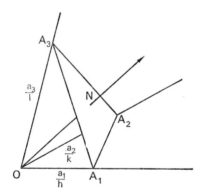

FIG. 7.8. The relation between sets of lattice planes and reciprocal lattice vectors.

7.5c. *Planes and vectors*

The reciprocal lattice vector $\mathbf{g} = h\mathbf{b}_1 + k\mathbf{b}_2 + l\mathbf{b}_3$ is normal to the set of lattice planes with Miller indices hkl and is of magnitude $|\mathbf{g}| = 1/d_{hkl}$, where d_{hkl} is the spacing of the lattice planes.

The proof is as follows: in Fig. 7.8, $A_1 A_2 A_3$ is part of the plane nearest to the origin of the set hkl, and $\overrightarrow{ON}$ is the normal to it from the origin: $|\overrightarrow{ON}| = d_{hkl}$. From the definition of Miller indices, $\overrightarrow{OA}_1 = \mathbf{a}_1/h$, $\overrightarrow{OA}_2 = \mathbf{a}_2/k$ and $\overrightarrow{OA}_3 = \mathbf{a}_3/l$.

Now $\qquad \overrightarrow{A_1A_2} = \overrightarrow{A_1O} + \overrightarrow{OA_2}$

$$= -\mathbf{a}_1/h + \mathbf{a}_2/k,$$

$$\therefore \quad \mathbf{g} \cdot \overrightarrow{A_1A_2} = (h\mathbf{b}_1 + k\mathbf{b}_2 + l\mathbf{b}_3) \cdot (-\mathbf{a}_1/h + \mathbf{a}_2/k).$$

$$= 0.$$

Therefore $\mathbf{g}$ is normal to $\overrightarrow{A_1A_2}$. Similarly, $\mathbf{g}$ is normal to $\overrightarrow{A_2A_3}$ and $\overrightarrow{A_3A_1}$, and so is normal to the plane.

To prove the second part we note that

$$d_{hkl} = |\overrightarrow{ON}| = \overrightarrow{OA_1} \cdot \frac{\mathbf{g}}{|\mathbf{g}|},$$

since $\mathbf{g}/|\mathbf{g}|$ is a unit vector normal to the plane $A_1A_2A_3$. Now

$$\overrightarrow{OA_1} \cdot \mathbf{g} = \frac{\mathbf{a}_1}{h} \cdot (h\mathbf{b}_1 + k\mathbf{b}_2 + l\mathbf{b}_3)$$

$$= 1,$$

$$d_{hkl} = 1/|\mathbf{g}|.$$

7.5d. *Non-lattice planes and points*

If we choose a primitive unit cell in the crystal lattice, then the unit cell in the reciprocal lattice will also be primitive, and to every set of numbers h, k and l there will correspond one reciprocal lattice point and one set of planes in the crystal lattice; this set of planes will be such that all the points of the crystal lattice lie on one or another plane of the set. Furthermore, the one-to-one relation between sets of lattice planes and reciprocal lattice points will hold regardless of the choice of unit cell. Now it is possible to visualise sets of lattice planes that do not include all the lattice points—for instance, the 100 planes of a face-centred cubic lattice when referred to the conventional unit cell. There will be no reciprocal lattice point corresponding to this set of planes, since all the reciprocal lattice points correspond to sets of planes that contain all the lattice points. Now if there is a lattice point a fraction $1/p$ (p integral) of the distance between a pair of planes from one of them, then, simply because it is a lattice point, there will be other lattice points at fractional distances of

$2/p$, $3/p$, ... $(p-1)/p$, and therefore the structure factor for this set of planes will be zero. This provides a method for detecting which of the points in a lattice reciprocal to, say, a primitive cubic lattice of cell edge a do not belong to the lattice reciprocal to the same lattice but with face-centring points added. This, in fact, is the usual method of detecting such extra points.

BIBLIOGRAPHY AND REFERENCES

BACON, G. E. (1955) *Neutron Diffraction*, O.U.P., Oxford.

BRADDICK, H. J. J. (1955) *Vibrations, Waves, and Diffraction*, McGraw-Hill, New York.

CULLITY, B. D. (1956) *Elements of X-ray Diffraction*, Addison-Wesley, Reading.

HIRSCH, P. B., HOWIE, A., NICHOLSON, R. B., PASHLEY, D. W. and WHELAN, M. J. (1965) *Electron Microscopy of Thin Crystals*, Butterworth, London.

JAMES, R. W. (1948) *The Optical Principles of the Diffraction of X-rays*, Bell, London.

JENKINS, F. A. and WHITE, H. E. (1950) *Fundamentals of Optics*, McGraw-Hill, New York.

LONGHURST, R. S. (1957) *Geometrical and Physical Optics*, Longmans, London.

INDEX

133